Applied Statistics and Econometrics

Bjørnar Karlsen Kivedal

Applied Statistics and Econometrics

Basic Topics and Tools with Gretl and R

palgrave
macmillan

Bjørnar Karlsen Kivedal
Østfold University College
Halden, Norway

ISBN 978-3-031-53141-5 ISBN 978-3-031-53142-2 (eBook)
https://doi.org/10.1007/978-3-031-53142-2

A previous edition of the book was published in Norwegian. "Anvendt statistikk og økonometri" by Bjørnar Karlsen Kivedal, © Bjørnar Karlsen Kivedal 2023. Published by H. Aschehoug & Co. (W. Nygaard) AS by Universitetsforlaget. All Rights Reserved.

Cover illustration: eStudioCalamar

This Palgrave Macmillan imprint is published by the registered company Springer Nature Switzerland AG
The registered company address is: Gewerbestrasse 11, 6330 Cham, Switzerland

Paper in this product is recyclable.

Preface

Applied statistics and econometrics can be complex and challenging for many students, but at the same time, it can be incredibly exciting, informative, and valuable once you unravel how to use the tools. As I have been teaching in this field for many years, I have noticed a significant need among students for an introduction to the fundamental tools in applied statistics and econometrics, with a strong focus on applications. This is the main reason why I came up with the idea of writing this book.

Primarily, this book is aimed at economics or business administration students. The examples are based on using Gretl or R as software, but the book is useful for learning the tools also if you are using other software or if you want to do manual calculations. The book can also be a relevant learning resource if you are a student in other disciplines, pursuing further education in quantitative methods, or in your professional career.

The topics in the book build upon the some foundations from basic mathematics and statistics courses. However, the most important basic elements from mathematics and statistics are also covered in the book to simplify the presentation. This makes the material accessible even if you do not necessarily have those prerequisites.

The book has its own website: www.palgrave.com. Here, you can find relevant exercises with solutions for each chapter, as well as datasets and script files with code used in the book. If you come across errors or have comments about the book, I appreciate it if you get in touch through the book's website.

I want to extend my thanks to colleagues and students who have provided input and inspiration over the years while I have been teaching the topics covered in the book. This book was initially published in Norwegian.

I would like to thank Associate Professor Johannes Mauritzen at NTNU Business School, who took the time read through the Norwegian manuscript to offer academic insights, as well as the editors Stian Brekke and Maria Almaas at Universitetsforlaget, who were immensely helpful throughout the work with the Norwegian version. Comments from Olvar Bergland at the Norwegian Univiersity of Life Sciences on the Norwegian version of the book has also been very helpful and improved the current version of the book. I would also like to thank the anonymous reviewers for feedback, and Ellie Duncan and the team at Palgrave Macmillan for all of the provided assistance during the process. Finally, I want to thank my family, colleagues and friends for their support throughout the entire process, including proofreading earlier drafts.

Halden, Norway
November 2023

Bjørnar Karlsen Kivedal

Contents

1 Introduction 1

1.1 Structure of the Book 1

1.2 Why Gretl and R? 2

2 Start Using Gretl and R 5

2.1 Gretl 5

2.1.1 Installing Gretl 6

2.1.2 Startup and Loading of Datasets 6

2.1.3 Data Visualization 8

2.1.4 Datasets 9

2.2 R and RStudio 12

2.2.1 Installation of R and RStudio 12

2.2.2 Startup and Loading Datasets 14

2.2.3 Some Simple R Codes 15

2.2.4 Loading Datasets 16

2.2.5 Data Visualization 17

3 Basic Material 19

3.1 Applied Statistics and Econometrics 19

3.2 Prerequisites 20

3.2.1 Summation 21

3.2.2 Derivation 21

3.2.3 Expectations 23

3.3 Different Summary Statistics 24

3.3.1 Central Tendency and Dispersion 24

3.3.2 Interval Estimate: Confidence Interval 27

3.3.3 Measures of Covariation: Covariance and Correlation 28

3.4 Probability Distributions 31

3.4.1 Normal Distribution 31

3.4.2 *Standard Normal Distribution* 32
3.4.3 *Central Limit Theorem* 33
3.4.4 *t-distribution* 34
3.4.5 *Probability Distributions in Gretl* 35
3.4.6 *Probability Distributions in R* 35
References 38

4 **Hypothesis Testing** 39
4.1 *Formulating Hypotheses* 40
4.2 *Test Statistic and Critical Value(s)* 40
4.2.1 *Example: Height of 9 Men—Normal Distribution* 42
4.3 *Examples in Gretl and R* 43
4.3.1 *Height of 9 Men:* t-*Distribution* 43
4.3.2 *Example Batteries: "Less Than" Alternative Hypothesis* 46
4.3.3 *Example with a Two-Sided Hypothesis Test* 52
4.4 *Hypothesis Testing for Correlation* 56
4.5 p-*Value* 56

5 **Simple Linear Regression** 59
5.1 *The Regression Line* 60
5.1.1 *A Straight Line: Math* 60
5.1.2 *A Straight Line: Regression* 62
5.1.3 *Interpreting the Estimated Coefficients* 64
5.2 *Simple Regression in Gretl* 66
5.3 *Simple Regression in R* 68
5.4 *Prediction* 69
5.5 *Explanatory Power of the Model* 71
5.6 *Hypothesis Testing and Confidence Intervals* 76
5.6.1 *Hypothesis Testing of the Slope Coefficient* 76
5.6.2 *Confidence Interval for a Slope Coefficient* 78
5.6.3 *Hypothesis Test for the Constant Term* 79
5.6.4 *Prediction Interval* 81
5.7 *Assumptions for the Regression Model* 84
5.7.1 *Graphical Assessment of the Assumptions* 85
5.8 *Supplementary Material* 88
5.8.1 *Where Do the Formulas for the Coefficients Come from?* 88
5.8.2 *Calculating the Numbers from a Regression Output* 89
Reference 93

6 **Multiple Regression** 95
6.1 *Manual Calculation* 96
6.2 *Interpreting Estimated Parameters in Multiple Regression* 97
6.2.1 *Partial Derivatives (Mathematics)* 97

6.2.2 *Interpretation If We Have Two Explanatory Variables* 98
6.3 *Example (in Gretl and R)* 99
6.3.1 *Example: Used Car Price and Fuel Consumption* 103
6.4 *F-Test* 106
6.4.1 *Example F-Test* 106
6.5 *Heteroskedasticity Tests* 110
6.5.1 *The Example in Gretl* 111
6.5.2 *Example in R* 112
6.5.3 *Dealing with Heteroskedasticity* 113
6.6 *Multicollinearity* 114
6.7 *Instrumental Variable Regression* 114
6.7.1 *Example Using an Instrumental Variable* 116
6.8 *Omitted Variable Bias* 121
References 122

7 **Regression Using Dummy Variables** 125
7.1 *Simple Regression with a Dummy Variable* 126
7.2 *Generate Dummy Variables in Gretl and R* 129
7.3 *Fixed Effects* 131
7.3.1 *Example: Wage, Education, and Sector* 132
7.3.2 *Hypothesis Testing* 134
7.4 *Interaction Effects* 134
7.4.1 *Example: Wage, Education, and Sector* 135
7.4.2 *Hypothesis Testing* 136
7.5 *Both Fixed Effects and Interaction Terms* 137
7.5.1 *Example* 138
7.5.2 *Hypothesis Testing* 140
7.5.3 *Example with Wage and City* 141
7.6 *Dummy Variable as Dependent Variable* 144
7.6.1 *Linear Probability Models* 144
7.6.2 *Probit Models* 145
7.6.3 *Logit Models* 151
7.6.4 *Which Probability Model to Use?* 155

8 **Non-Linear Relationships** 157
8.1 *Logarithmic Relationships* 161
8.1.1 *Lin-Log Model* 161
8.1.2 *Log-Lin Model* 164
8.1.3 *Log-Log Model* 165
8.1.4 *Predicted Y when* $\ln Y$ *Is the Dependent Variable* 165
8.2 *Polynomial Functions* 166
8.3 *Inverse Relationships* 170
8.4 *Which Non-Linear Model Should We Choose?* 172
8.5 *Supplementary Material* 174

8.5.1 *Logarithms and Approximation of Percentage Change* 174
8.5.2 *Lin-Log Regression by Hand* 175
Reference 176

9 Time Series Data 177
9.1 *Time Series Variables and Regressions* 178
9.1.1 *Time Series in Gretl and R* 178
9.1.2 *Regression with Time Series* 180
9.1.3 *Development Over Time: Trend* 181
9.1.4 *Spurious Regression* 184
9.1.5 *Transforming Time Series Variables* 187
9.2 *Autocorrelation and Stationarity* 191
9.2.1 *Autoregressive Model* 191
9.2.2 *Autoregressive Model with Multiple Lags* 195
9.2.3 *AR Models in Gretl and R* 195
9.2.4 *AR Models in Gretl and R* 196
9.2.5 *Stationarity Test* 197
9.3 *Seasonal Variation* 203
9.3.1 *Seasonal Variation in Gretl and R* 204
9.3.2 *AR(1) Model with Drift and Seasonal Variation* 206
9.4 *Forecasts and Model Evaluation* 206
9.4.1 *Evaluating Forecasts in Gretl and R* 210
9.5 *Testing for Autocorrelation in the Residuals* 211
9.5.1 *Testing for Autocorrelated Residuals in Gretl and R* 212
9.6 *Parameter Stability* 214
9.6.1 *Chow Test in Gretl* 215
9.6.2 *Chow Test in R* 216
9.7 *Cointegration* 217
9.7.1 *Cointegration in Gretl* 218
9.7.2 *Cointegration in R* 221
References 223

10 Other Statistical Tools 225
10.1 *T-test* 225
10.1.1 *Example* 226
10.1.2 *T-test or Regression Model* 229
10.2 *Chi-square Test* 230
10.2.1 *Chi-square Goodness of Fit Test* 231
10.2.2 *Gretl and R* 232
10.2.3 *Example with Data* 233
10.2.4 *Chi-square Test or Regression Analysis* 236

10.3 Analysis of Variance 238
10.3.1 Example in Gretl 240
10.3.2 Example in R 241
10.3.3 Analysis of Variance or Using Dummy Variables 242

Bibliography 243

Index 245

List of Figures

Fig. 2.1	Website for Gretl	6
Fig. 2.2	Startup window of Gretl.eps	7
Fig. 2.3	Manual data entry in Gretl	8
Fig. 2.4	Open/Import other datasets	9
Fig. 2.5	Main window with data	10
Fig. 2.6	Display of all observations	10
Fig. 2.7	Exporting data	11
Fig. 2.8	Toolbar in the main window	11
Fig. 2.9	Scatter plot between wage and education	12
Fig. 2.10	Editing scatter plots	13
Fig. 2.11	Available datasets in the "Gretl" Tab	13
Fig. 2.12	RStudio	14
Fig. 2.13	RStudio with script window	15
Fig. 2.14	Simple operation in R.eps	16
Fig. 2.15	Scatter plot	18
Fig. 2.16	Scatter plot	18
Fig. 3.1	Statistical inference	20
Fig. 3.2	The polynomial function $y = 2 + 0.5x + 0.1x^2$ and its tangents for $x = 1$ and $x = 3$, with slope of 0.7 and 1.1, respectively	22
Fig. 3.3	Central tendency and dispersion in R	27
Fig. 3.4	Correlation matrix for two variables	30
Fig. 3.5	Normal distribution. The height of 18-year-old men	32
Fig. 3.6	Standard normal distribution	33
Fig. 3.7	Normal distribution, and the t-distribution with 2, 5, and 20 degrees of freedom	35
Fig. 3.8	Statistical tables in Gretl	36
Fig. 4.1	Type 1 and Type 2 errors	41
Fig. 4.2	The Normal distribution with critical values at a 5% significance level. Red areas show the rejection area	42
Fig. 4.3	Standard normal distribution with vertical lines representing the test statistic $Z = 0.647$ and the critical value $cv = 1.645$	43
Fig. 4.4	Outcome of hypothesis testing	45

Fig. 4.5	Standard normal distribution with a critical value of $cv = -1.645$ and a test statistic of $Z = -1.936$	47
Fig. 4.6	Descriptive statistics of battery lifetimes	48
Fig. 4.7	Battery Hypothesis Test in Gretl, Known Population Standard Deviation	50
Fig. 4.8	Battery Hypothesis Test in Gretl, Unknown Population Standard Deviation	51
Fig. 4.9	The standard normal distribution with vertical lines for the test statistic $Z = 0.647$ and the critical value $cv = 1.645$	57
Fig. 5.1	Scatterplot between wage and education	60
Fig. 5.2	A straight line with a slope coefficient of 0.5 and intersection of 1.0 ($y = 0.5x + 1$)	61
Fig. 5.3	Deviation and squared deviation between observations (Y_i^*) and the regression line $\hat{Y}_i^* = 1.123 + 0.719 \cdot X$, where $Y_i^* = Y/100$ in order to make the figure easier to read	63
Fig. 5.4	Regression line	65
Fig. 5.5	Model specification	66
Fig. 5.6	Simple regression Gretl output	67
Fig. 5.7	Simple regression in Gretl console	68
Fig. 5.8	Regression line and residual	70
Fig. 5.9	Analysis of variance (ANOVA) in Gretl	74
Fig. 5.10	Confidence interval for coefficients in Gretl	79
Fig. 5.11	Prediction interval in Gretl	82
Fig. 5.12	Prediction interval output in Gretl	82
Fig. 5.13	Prediction interval, figure	83
Fig. 5.14	Graphical presentation of residuals and predicted values in Gretl	86
Fig. 5.15	Graphical presentation of residuals and predicted values in R	87
Fig. 6.1	Multiple regression, variable selection	100
Fig. 6.2	F-distribution with two degrees of freedom in the numerator and five in the denominator; $F(2, 5)$	108
Fig. 6.3	Scatterplot between the explanatory variable X_4 and residual $\hat{u}$	111
Fig. 6.4	Two-stage least squares in Gretl	118
Fig. 7.1	Generate dummy variables in Gretl	130
Fig. 7.2	Regression lines when using a fixed effect	133
Fig. 7.3	Regression lines when using an interaction term	137
Fig. 7.4	Regression lines using both fixed and interaction effects	139
Fig. 7.5	Linear Probability Model	145
Fig. 7.6	Cumulative standard normal distribution	146
Fig. 7.7	Observations and predictions from the estimated probit model	148
Fig. 8.1	Wage, experience, and regression line	159
Fig. 8.2	Residuals and experience	159
Fig. 8.3	Wage and experience. Linear function and logarithmic function	160
Fig. 8.4	Wage, log experience, and residuals	162
Fig. 8.5	Wage and experience. Linear function and second-degree polynomial function	169
Fig. 8.6	Wage and experience. Linear function, quadratic function, and cubic function	170
Fig. 9.1	Time series data in Gretl	179

Fig. 9.2 Scatterplot of sales and price 180
Fig. 9.3 Time series plot of quantity sold and price 182
Fig. 9.4 Time series plot of quantity sold and price created in R 182
Fig. 9.5 Scatterplot and regression line for the total motorway accident fatality rate in the USA (Y_t) and the number of tons of fresh lemons imported from Mexico to the USA (X_t) 184
Fig. 9.6 Time series of the total motorway accident fatality rate in the USA (Y_t) and the number of tons of fresh lemons imported from Mexico to the USA (X_t) 186
Fig. 9.7 First difference of sales and price 189
Fig. 9.8 Plot of Z_t and W_t 192
Fig. 9.9 Quarterly data for sales first quarter 2000 ($t = 1$) to fourth quarter 2002 ($t = 12$) 193
Fig. 9.10 Lagged terms in Gretl 196
Fig. 9.11 Augmented Dickey-Fuller test in Gretl 200
Fig. 9.12 Dummy variables for season, variable list, and dataset 205
Fig. 9.13 Sales and price 2001–2012 214
Fig. 9.14 Residuals from OLS regression on $\text{sales}_t = \beta_0 + \beta_1 \text{price}_t + u_t$ 221
Fig. 10.1 Unpaired t-test. Figure from test in Gretl 229
Fig. 10.2 Data for chi-squared test in Gretl 233
Fig. 10.3 Chi-squared test in Gretl 235

List of Tables

Table 2.1	Wage and education for eight individuals	7
Table 3.1	Calculation of the correlation coefficient between X (education) and Y (wage)	29
Table 3.2	Dice rolls	34
Table 4.1	Battery lifetimes	48
Table 4.2	Wages in 2021 (wage21) and 2022 (wage22)	52
Table 4.3	Wage growth in euros	52
Table 5.1	Calculation of the correlation coefficient	64
Table 5.2	Observations and predicted dependent variable	71
Table 5.3	Predicted dependent variable, error term, and squared error term	72
Table 5.4	Predicted dependent variable and error terms	92
Table 6.1	Dataset: wage, education, and experience	99
Table 6.2	Results	105
Table 6.3	Dataset: wages, education, and distance to nearest college	116
Table 7.1	Wage, education, experience, sector, and gender	126
Table 7.2	Wage, education, experience, sector, and country	131
Table 7.3	Dataset with generated interaction term	135
Table 8.1	Wage and experience	158
Table 8.2	Wage, experience, and residuals	160
Table 8.3	Wage, log experience, and residuals	162
Table 8.4	Wage and number of sick days	171
Table 8.5	Calculations for estimating coefficients in a lin-log model	176
Table 9.1	Sold quantity and price six years, 2001–2006	178
Table 9.2	Total motorway accident fatality rate in the USA (Y_t) and the number of tons of fresh lemons imported from Mexico to the USA (X_t)	185
Table 9.3	Lagged variables	187
Table 9.4	Sold quantity and price. Level and change	188
Table 9.5	Growth in sales	190
Table 9.6	Quarterly data for sales, changes (dsales), and growth (salesgrowth)	194

Table 9.7	Actual sales and predicted sales with AR(1) model (1), AR(1) model with drift (2), and AR(1) model with drift and dummy variable for the third quarter (3)	208
Table 9.8	Residuals for AR(1) model (1), AR(1) model with drift (2) and AR(1) model with drift and dummy variable for the third quarter (3)	208
Table 9.9	Calculations for forecast errors of the AR(1) model	209
Table 10.1	Wage for employees in the public and private sectors	226
Table 10.2	Wage in the public and private sectors	227
Table 10.3	Adults and children buying products A and B	230
Table 10.4	Expected values assuming children and adults have the same distribution	231
Table 10.5	Squared deviations relative to expected value	231
Table 10.6	Wages and Country of Residence	238
Table 10.7	Wages and City of Residence	238

CHAPTER 1

Introduction

This chapter presents the structure of the book and the motivation and background for focusing on Gretl and R in a book on applying basic topics and tools in statistics and econometrics.

1.1 Structure of the Book

This introductory chapter provides an overview of, and background for, the content in the book. In the next chapter, we will start to look at Gretl and R, with a focus on installation and loading datasets, as well as some simple tools for descriptive statistics. I will then include the use of Gretl and R in each chapter in connection to the topics presented, to demonstrate how we can perform various analyses.

Each chapter thus covers different tools, with Chapter 3 starting with an overview of relevant basic statistics, while we proceed with hypothesis testing in Chapter 4.

In Chapter 5, we cover regression analysis and how this can be done in Gretl and R. For those with a certain basic understanding of statistics who want to learn regression analysis and/or the use of Gretl and R for regression analysis, it's possible to go directly to Chapter 5 after installing and starting Gretl and/or R as described in Chapter 2.

We expand the simple regression model to include multiple explanatory variables in Chapter 6. This does not lead to significant changes in the execution within Gretl or R, but it does require interpreting the estimates in a new way since we include control variables or instrumental variables. This is important for trying to isolate the effects we want to examine. If you already

B. Karlsen Kivedal, *Applied Statistics and Econometrics*,
https://doi.org/10.1007/978-3-031-53142-2_1

have a good understanding of simple regression, you can proceed directly to Chapter 6.

The use of dummy variables is introduced in Chapter 7, allowing us to distinguish between different groups in the data. In this chapter, we will calculate fixed effects and interaction effects, in addition to estimating probability models for various outcomes.

Regression models can also be used to estimate non-linear relationships. This can be done by transforming the variables through logarithmic or polynomial functions. Different methods for doing this are presented in Chapter 8.

We are often interested in analyzing data observed over time, such as stock prices, sales or demand, or macroeconomic variables. In time series analysis, we encounter some new challenges that we should investigate before using the regression model. We will explore some fundamental topics and models in time series analysis in Chapter 9.

In Chapter 10, we examine some other tools commonly used in applied statistics: t-tests, chi-square tests, and analysis of variance (ANOVA). These are not extensions of the regression model, but as we will see in this chapter, it is possible to perform similar analyses with the regression model. This chapter can be read without prior knowledge of the regression model, directly after Chapters 3 and 4.

First, we briefly explore the theoretical aspects of the various topics to provide the necessary background and context. In most cases, we will use small examples with manual estimation in addition to analysis using Gretl and R. This approach makes the datasets and software outputs more transparent, allowing for a better understanding of all observations and variables before working with larger datasets in exercises, such as those found in the book's online resources.

1.2 Why Gretl and R?

After teaching a second year bachelor course in Applied Statistics for students in business administration over many years, it has been challenging to find a suitable textbook in applied statistics and econometrics. This is one of the reasons why I decided to write this book. Not just for me, but also for you as a student. It is important to be introduced to relevant analysis and data processing tools that can be used for, for example, a bachelor's thesis or for other projects. I would like to point out that the content of the book is not at the same advanced level as the textbooks that are usually used in econometrics and statistics subjects in master's degree studies.

This book emphasizes using Gretl and R to do analysis and data processing. Both Gretl and R are free, readily available, open-source programs, and they are easy to use for basic applied statistics and econometrics. It is therefore possible to follow the book whether you want to use R or Gretl. I also mainly do the analyses in the book by hand with small examples, to show you what the programs actually do.

There is not much available literature about applying Gretl, so I believe that it is important to base parts of the book around this software. Gretl is often highlighted as statistical software that fits well in a teaching context. In addition, Gretl can be used for more advanced econometrics and data processing than what is covered in this book. Microsoft Excel can be used for many of the topics covered in the book, but since Gretl is designed for applied statistics and econometrics, it is much easier and faster to use this for such analyses. It is also possible to import commands from other programming languages such as R, Python, Julia, Ox, and Stata.

Although there are already several books that describe many of R's functionalities, I found it to be equally important that this book also deals with this data processing tool, since R as software is already widely used at many educational institutions and workplaces.

The book's core theme, alongside applications of the aforementioned software, is regression analysis. The combination of regression analysis and other statistical tools in this book means that it both covers advanced topics in other introductory statistics books that are often used in business administration studies, and it provides a more comprehensive introduction to regression analysis and econometrics than what is usually found there.

CHAPTER 2

Start Using Gretl and R

In this chapter we look at how to get started with Gretl and R. In addition to a short introduction to opening the programs and loading datasets, we also perform some basic summary statistics and do some data visualization.

2.1 Gretl

Gretl (Gnu Regression, Econometrics and Time-series Library) is open-source software, which means you can use the program independently of any purchasing agreements your college, university, or employer might have. Most likely, you do have access to Microsoft Excel, where you can perform many of the tasks presented with Gretl in this book. However, Excel is best suited for very basic tasks. Additionally, Gretl is specifically designed for econometrics and applied statistics, so it is more convenient and often faster to use Gretl even for simple tasks. The syntax will also differ in Excel depending on which language you use in the interface of the program. This can make working with Excel more challenging if you for example have a different language version of Excel installed than your lecturer or colleagues. The automatic formatting in Excel can also pose problems when analyzing datasets.

The dataset used for the regression analysis, which constitutes a significant portion of the book, is a small constructed dataset with only eight observations. This allows us not only to perform analyses in Gretl but also to calculate many things manually, making it easier to understand the relationship between the dataset and what Gretl actually does.

B. Karlsen Kivedal, *Applied Statistics and Econometrics*,
https://doi.org/10.1007/978-3-031-53142-2_2

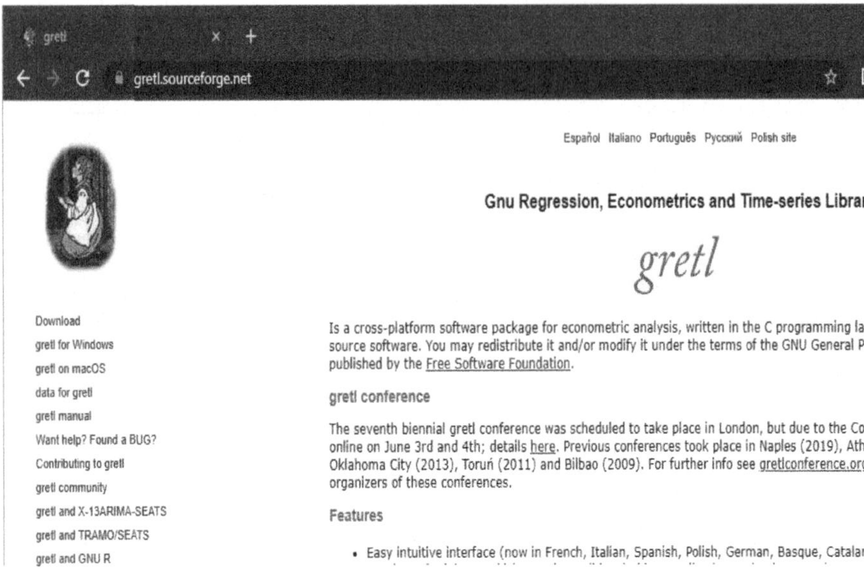

Fig. 2.1 Website for Gretl

2.1.1 Installing Gretl

Gretl is free and open-source, and it can be downloaded from http://gretl.sourceforge.net/. There are versions available for Windows, Mac OS X, and Linux. The screenshots and descriptions in this book are taken from Windows, but the program works in the same way for other operating systems (Fig. 2.1).

2.1.2 Startup and Loading of Datasets

After installing Gretl and opening the program, it will look like shown in Fig. 2.2.

This will be referred to as the **main window** in Gretl because you will also see many other windows for various calculations and estimations. In this main window, we will have an overview of the variables in our dataset. If we have a sample of eight individuals, with $i = 1, \ldots, 8$, where we have information on weekly income measured in Euros Y_i and years of education after high school X_i. This results in eight pairs of observations (Table 2.1).

The first person in the sample ($i = 1$) thus earns 320 Euros and has two years of education after high school, while the fifth person has five years of education after high school and earns 510 EUR.

You can load data into Gretl in several ways. For small datasets or if we collect data manually, we can enter the data directly into Gretl by choosing `File > New dataset` in the main window (Fig. 2.3).

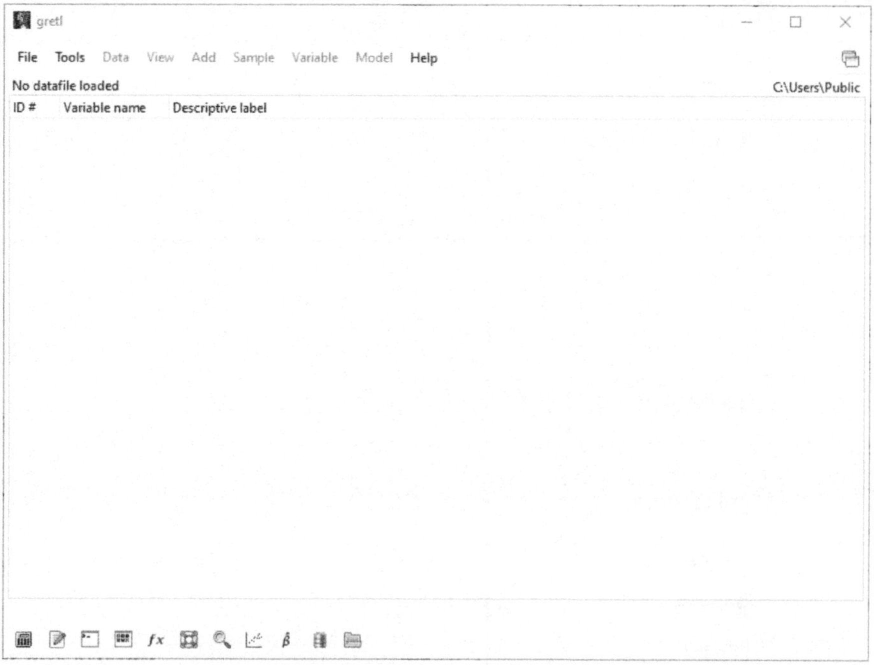

Fig. 2.2 Startup window of Gretl.eps

Table 2.1 Wage and education for eight individuals

i	*1*	*2*	*3*	*4*	*5*	*6*	*7*	*8*
X_i	2	3	1	5	5	4	7	5
Y_i	320	290	170	480	510	410	650	370

We can open the data we have in a file, such as a gdt file (Gretl dataset), an Excel file (xls or xlsx), or from other supported file formats. This can be done via `File > Open data > User file...`. After that, we need to select the file format if we are opening something other than a gdt file (Fig. 2.4).

After the data (the two variables, wage, and education) are loaded, the main window in Gretl looks like shown in Fig. 2.5.

By clicking on one of the variables (or selecting both), we can go to `Data > Display values` in the tool menu to see the observations as they appear in, for example, a spreadsheet such as Excel, as shown in Fig. 2.6. We can also change the name or description of the variable by going to `Variable > Edit attributes`.

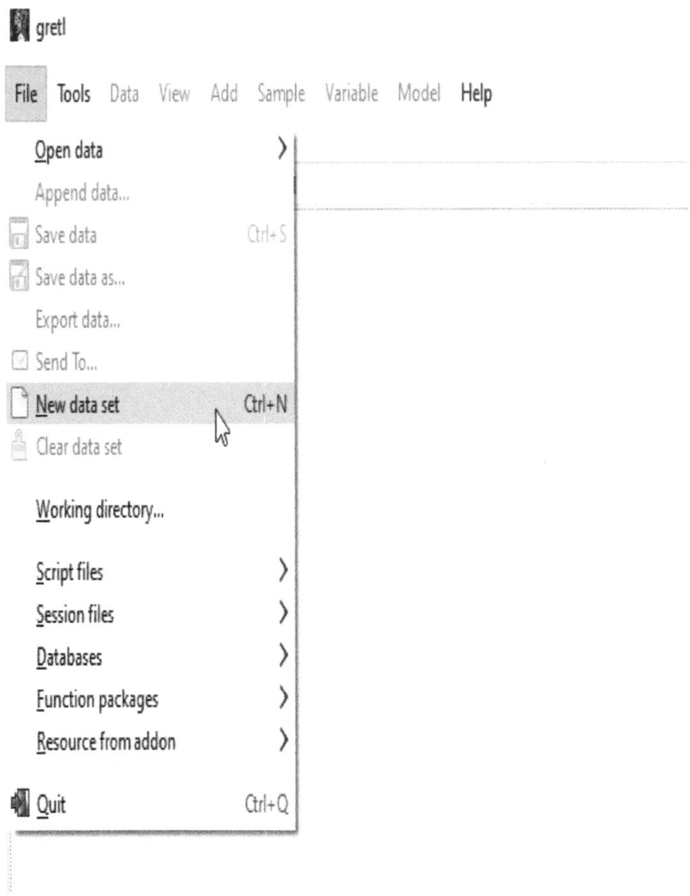

Fig. 2.3 Manual data entry in Gretl

Exporting Datasets
Additionally, we can export a dataset to a file that can be opened in Excel by selecting `File` > `Export data`. We can then choose the variables we want to save (`All ->` to include all) as shown in Fig. 2.7. We can for example select the CSV (Comma Separated Value) format, which can be opened directly in Excel or other programs. Then, we can choose the appropriate `separator for data columns`, the default being `comma (,)`.

2.1.3 Data Visualization

We can create a scatter plot by clicking on the highlighted button, which is the fourth from the right (X-Y graph) in the lower toolbar of the main window (Fig. 2.8).

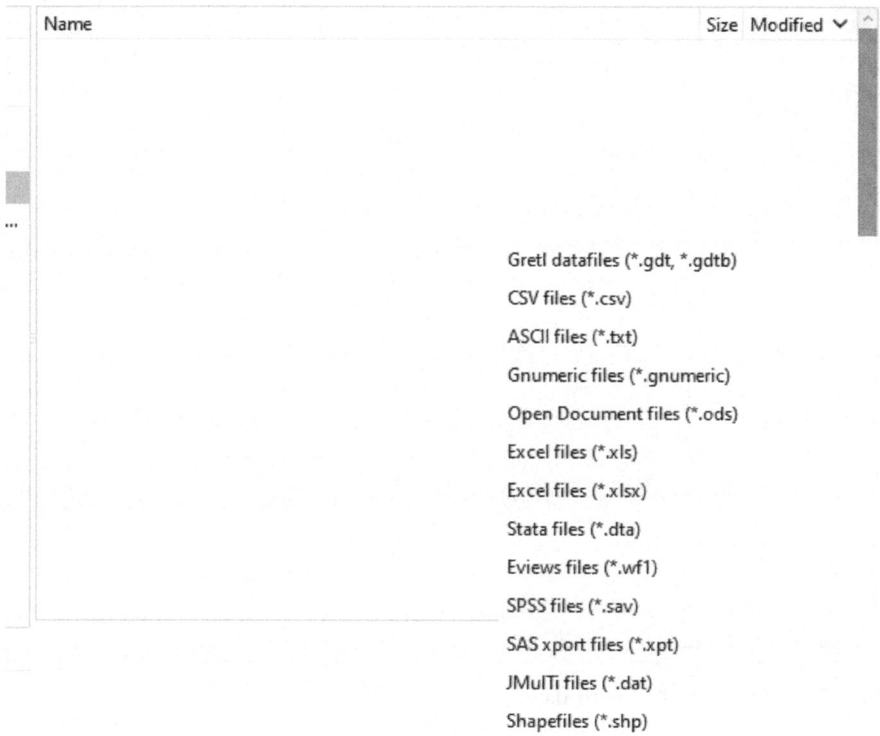

Fig. 2.4 Open/Import other datasets

There, we can choose education on the x-axis (the horizontal axis) and wage on the y-axis (the vertical axis). We can also choose `suppress fitted line`, which will remove the regression line (more on this in Chapter 5). This will open a window with the scatter plot between the two variables (Fig. 2.9).

In the scatter plot, the small circles represent each observation pair, and the straight line (if we did not choose to suppress it) is the regression line with its corresponding equation in the top left corner of the figure (more on this in Chapter 5). You can remove the regression line by right-clicking on the figure, selecting `Edit`, and choosing `none` under `fitted line` to obtain a clean scatter plot. You can also modify how the observations appear in this menu (Fig. 2.10).

This figure can be saved as an image or copied and pasted into a document. Right-clicking on the image will bring up the options.

2.1.4 Datasets

Through Gretl, you have access to many datasets that are used as sample data in various econometrics textbooks. By going to `Open data > Sample file...`, there are already some tabs indicating groups of available datasets

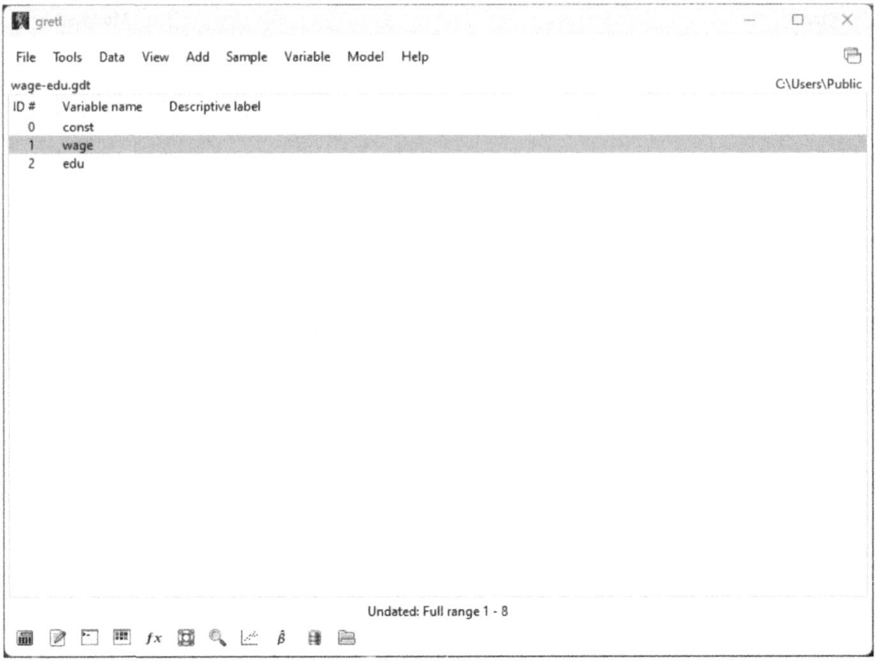

Fig. 2.5 Main window with data

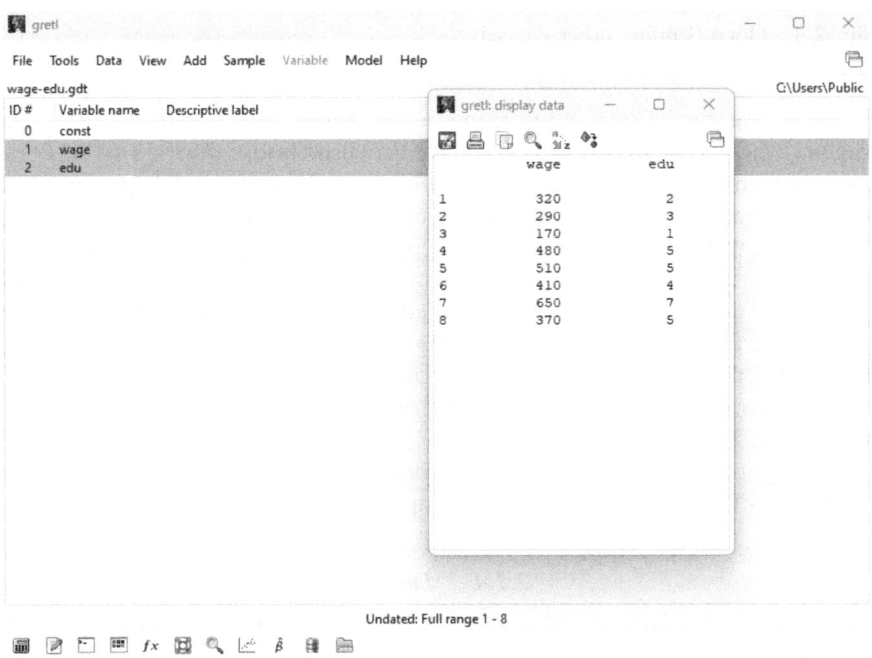

Fig. 2.6 Display of all observations

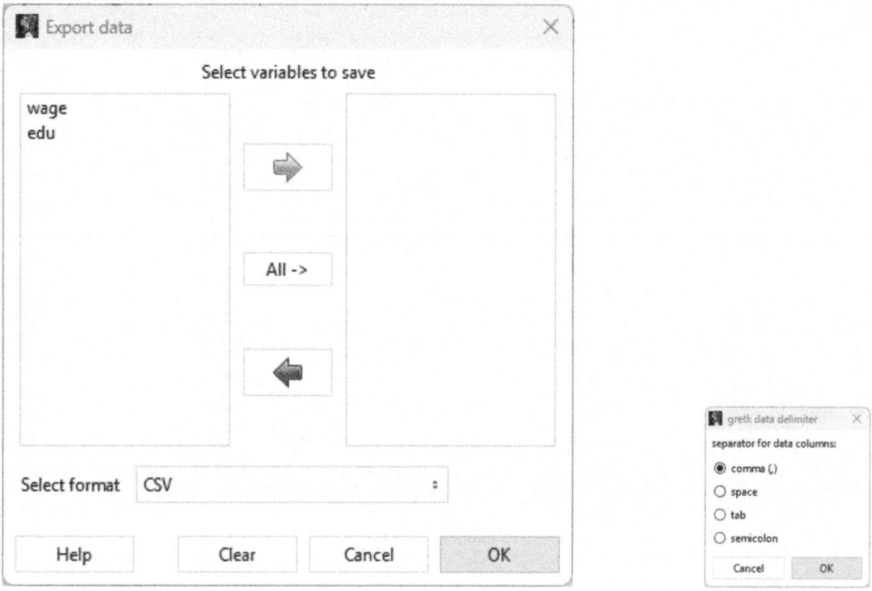

Fig. 2.7 Exporting data

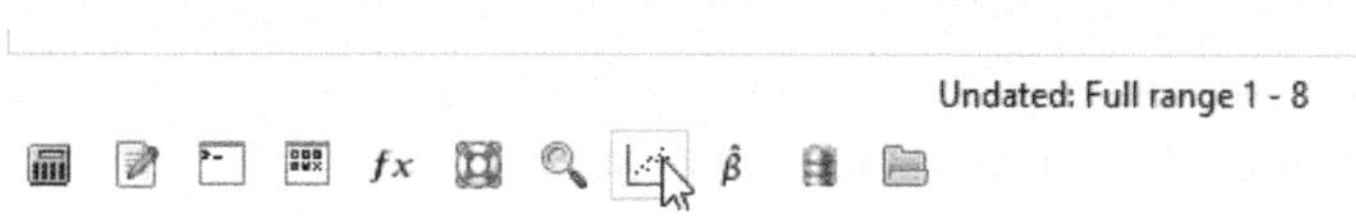

Fig. 2.8 Toolbar in the main window

(Gretl, Greene, and Koop). In the `Gretl` tab, there are various datasets known from different literature, as shown in Fig. 2.11, while `Greene` and `Koop` contain datasets used in the textbooks Econometric Analysis (Greene 2003) and Analysis of Economic Data (Koop 2013), respectively.

Additionally, you can click on `Look on server` (the third button in the window) to download datasets from other textbooks. This can be useful for replicating examples and working on exercises from these books. Visit https://gretl.sourceforge.net/gretl_data.html for an overview of what is available. These datasets are often also available on the respective textbooks' websites, but through the Gretl menu, you can open them directly in Gretl. Many of the exercises related to the chapters in this book are constructed based on data from these datasets.

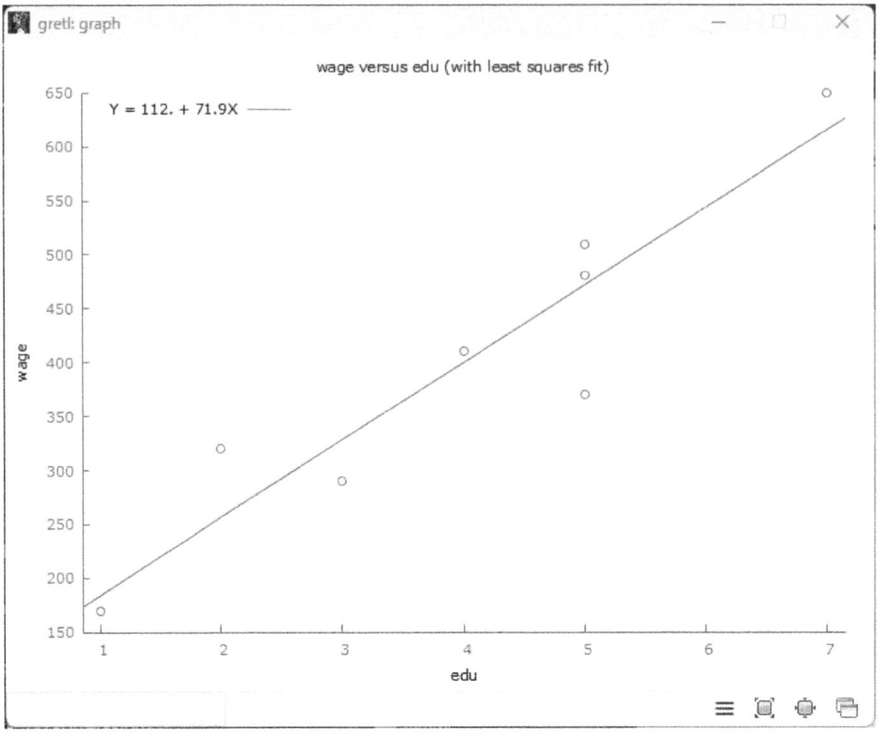

Fig. 2.9 Scatter plot between wage and education

2.2 R and RStudio

R, like Gretl, is free and open-source. While you can download R directly from, for instance, https://www.r-project.org/ and use R independently, it's common to use something like **RStudio**, created by Posit, as an interface for R. RStudio is not a graphical interface like Gretl, although it does have some menus that make it easier to use certain functions.

There is a lot of literature available on using R, covering both basic functions and more advanced analyses. However, the focus of this book is to provide a basic introduction to simple topics in applied statistics and econometrics. There are numerous ways to use R through various add-on packages and functions. I will try, as far as possible, to use the built-in functions in R that do not require the installation of additional packages (although this is also straightforward to do).

2.2.1 Installation of R and RStudio

To use RStudio, which will be the focus of this book, R must first be installed, as mentioned in the previous section. Afterward, RStudio can be downloaded

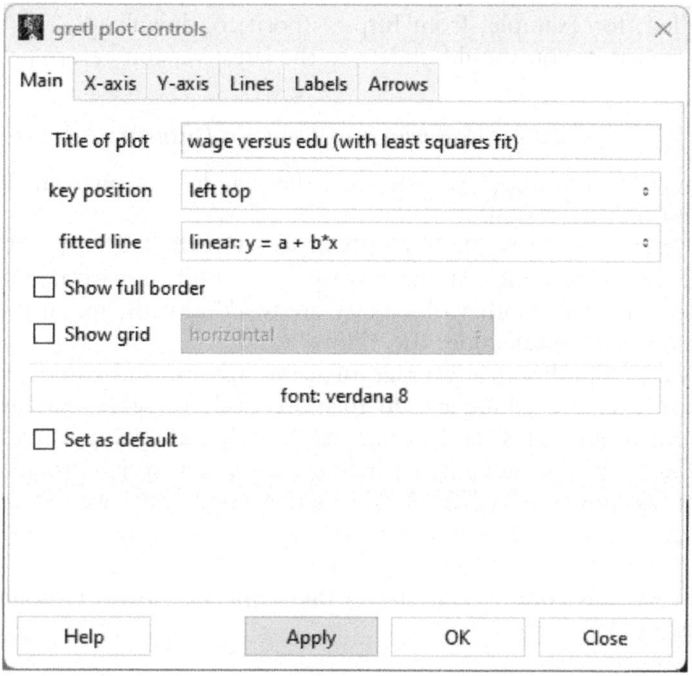

Fig. 2.10 Editing scatter plots

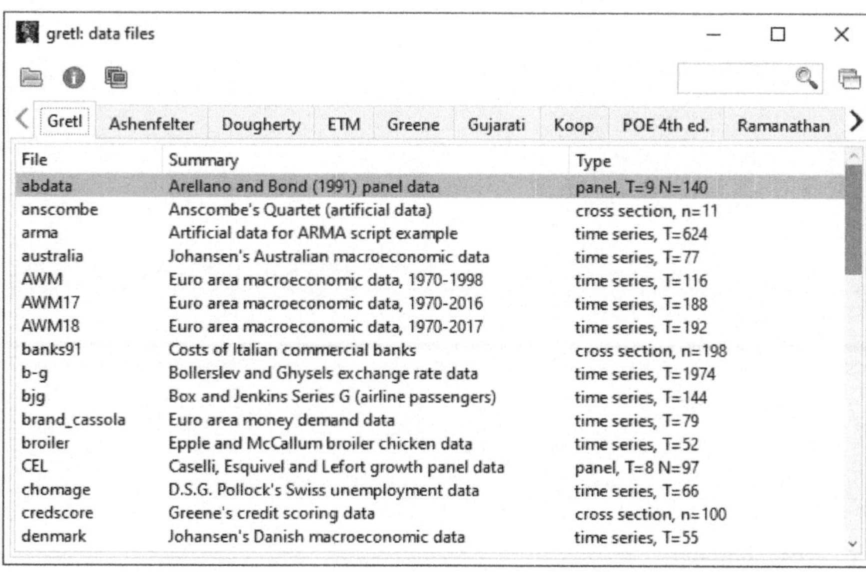

Fig. 2.11 Available datasets in the "Gretl" Tab

and installed, for example, from https://posit.co/downloads/. If you have already installed it, you should update to the latest version before proceeding.

2.2.2 *Startup and Loading Datasets*

After installing RStudio, we will see the window shown in Fig. 2.12 (depending on the version).

Upon startup, we have three panels in use. On the left is the "Console", where we can write code. At the top right, we find "Environment", which will display variables or other objects we are working with, and at the bottom right, figures will appear under the "Plots" tab.

However, a significant advantage of using a command-based program is that we can save everything we do to more easily preserve our history and replicate our results. This can be done with a **script**. By going to `File` > `New File` > `R script`, we get a new panel in the top left, which provides a blank document, as shown in Fig. 2.13. This is the script where we can enter code or commands, preferably one command per line. We can save this script by pressing, for example, `Ctrl+S` or go to `File` > `Save`.

We can write R code in RStudio in the script file (which typically has the file extension `.R`).

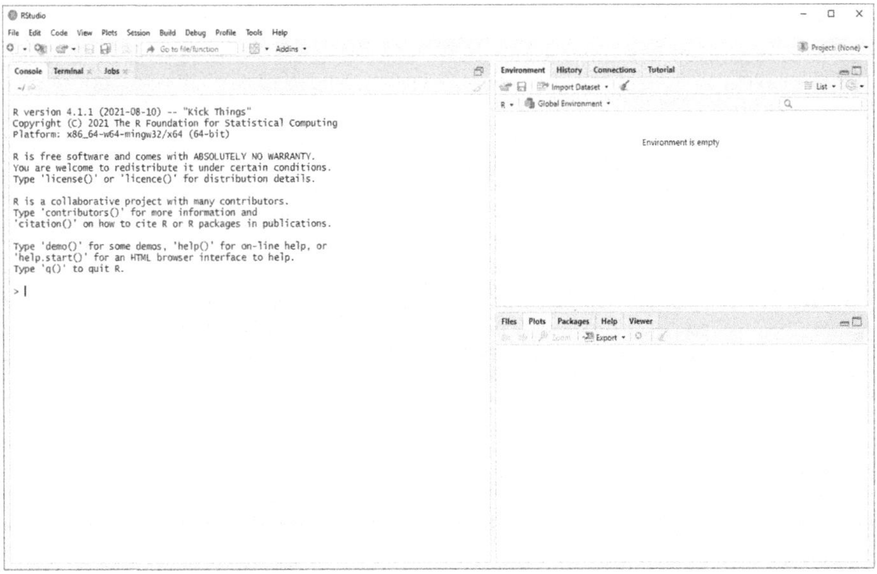

Fig. 2.12 RStudio

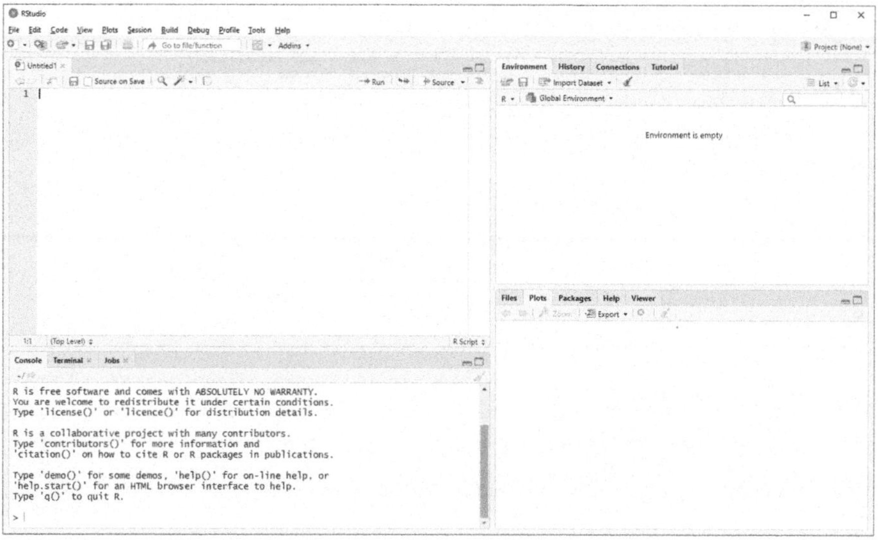

Fig. 2.13 RStudio with script window

2.2.3 Some Simple R Codes

There are several ways to use R, and here, we will demonstrate some simple calculations to illustrate how variables and objects are used in R, which we may need later. For example, we can define the variable `x` as 2 by writing `x < - 2` and use this to create the variable `y`, given by $y = 2x + 3$, which is $y = 7$ when $x = 2$, by writing `y <- 2*x+3`. We can write these as two lines in the script window. Then, we can click `Run` for each line, press `Ctrl+Enter` on each line, or select both lines and click `Run` or press `Ctrl+Enter`. We will then see that x is 2, and y is 7 in the "Values" panel, as shown in Fig. 2.14.

R code 2.1: Simple Operation in R

```
x <- 2
y <- 2*x+3
```

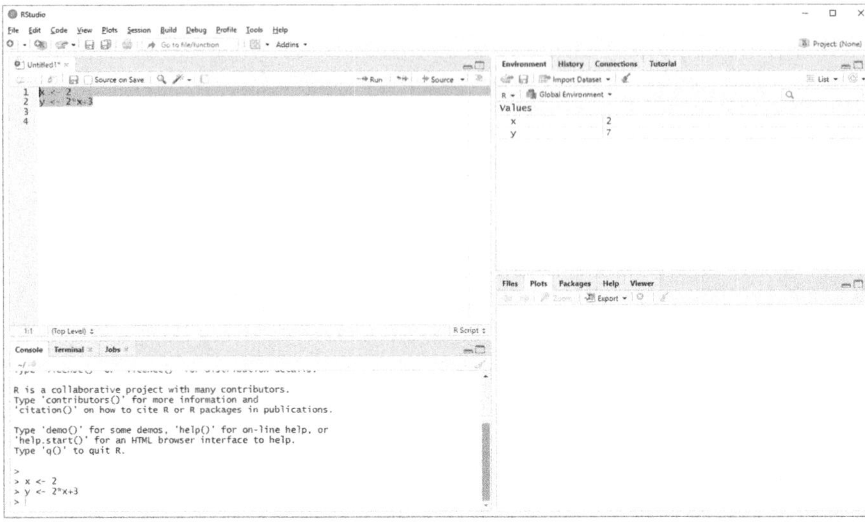

Fig. 2.14 Simple operation in R.eps

2.2.4 *Loading Datasets*

We can enter the observations for X and Y from Table directly into R.

i	*1*	*2*	*3*	*4*	*5*	*6*	*7*	*8*
X_i	2	3	1	5	5	4	7	5
Y_i	320	290	170	480	510	410	650	370

We can do this in the script window as shown in the R code below, where X_i is `education` and Y_i is `wage`. In other words, by defining a vector for our variable using `c()`, which contains the values the vector should have, separated by commas.

R code 2.2: Entering Data in R

```
wage <- c(320,290,170,480,510,410,650,370)
edu <- c(2,3,1,5,5,4,7,5)
```

We can also import data from other files into R. The easiest way is to import data from a csv file (e.g., saved in Excel). However, be aware that Excel may have automatically formatted some cells, so it is a good idea to check the csv file in, for example, Notepad in Windows or another plain text program to ensure that everything looks correct. First, we must define a working directory on our local PC, e.g., `C:\Users\username\Rdata` and then we can load the csv file from this directory or a subdirectory using the `read.table` function, as shown in the R code below. In the line for importing data below, it is assumed that a comma separates the columns in the csv file. `fill=TRUE` means that any empty cells will be imported if each column is of different lengths, and `header=TRUE` means that the first line in the csv file contains the names of each variable. For an overview, you can type `help(read.table)` in the console window.

R code 2.3: Loading Data from a CSV File in R

```
setwd("C:/Users/username/Rdata")
wagedata <- read.table(file="wageedu.csv", sep=",",
          fill=TRUE, header = TRUE)
```

2.2.5 Data Visualization

We can easily create a scatter plot between `edu` and `wage` using the `plot` function, as shown in the box below.

R code 2.4: Scatter Plot in R

```
plot(edu, wage, main="Scatter Plot of Wage and Education",
    xlab="edu", ylab="wage", pch=19)
```

This will generate the plot as seen in Fig. 2.15. You can also save it by clicking "Export" in the plot window. Here, `main` refers to the title, `xlab` is the label for the x-axis, and `ylab` is the label for the y-axis. `pch` specifies the style of the points, where 19 corresponds to solid black points as shown in the figure. You can refer to `help(pch)` for more details. The exported figure is presented in Fig. 2.16.

There are also numerous built-in datasets available in R. You can list them by using `data()`. To load a dataset, simply type `data(datasetname)`, for example, `data(uspop)`, which provides a time series with observations for every tenth year on the population in the United States from 1790 to 1970, measured in millions of people (more on time series in Chapter 9).

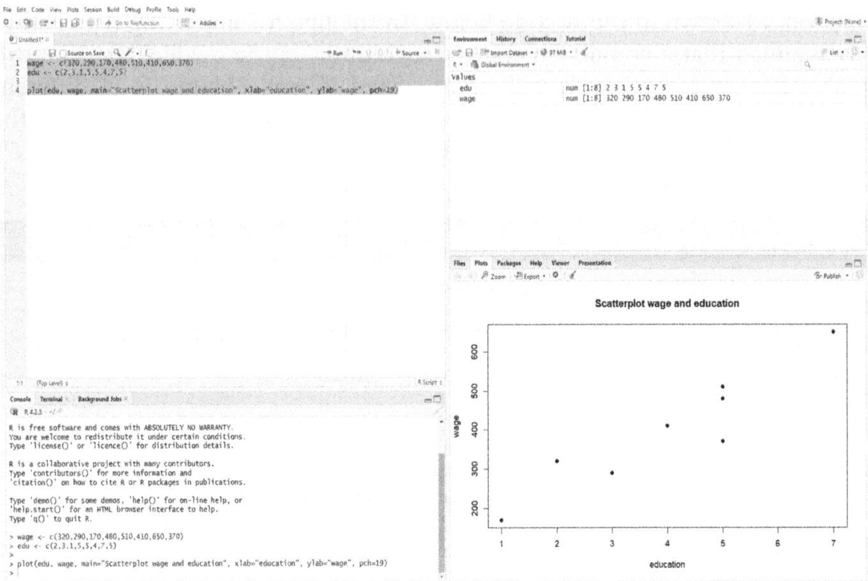

Fig. 2.15 Scatter plot

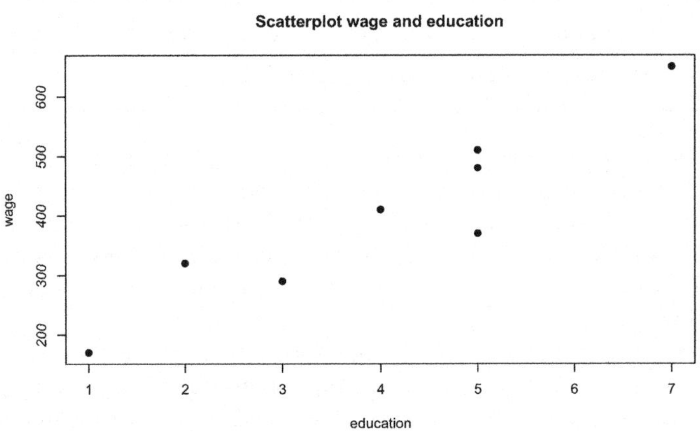

Fig. 2.16 Scatter plot

CHAPTER 3

Basic Material

This chapter covers elementary topics in statistics which are necessary to comprehend before going to the next chapters. The main topic is estimation, looking at both point and interval estimates. We also look at correlation analysis as the first step of investigating relationships between variables which is further elaborated in the next chapters. The most used probability distributions are also introduced here, both from a theoretical standpoint and related to Gretl and R.

3.1 Applied Statistics and Econometrics

Within statistics, we often distinguish between descriptive statistics and statistical inference. Descriptive statistics primarily involves summarizing data through various measures of central tendency, dispersion, and covariation. Common quantities that are then calculated are mean, median, mode, variance, standard deviation, covariance, and correlation. Graphical presentation of the data (data visualization) is also a form of descriptive statistics.

In statistical inference, we use a sample that provides us with data to say something about the population. It is then important that the sample provides a representative picture of the population, and we are aware of the uncertainty in our calculations. As illustrated in Fig. 3.1, one often draws a sample consisting of a group from a larger population. This sample is then used to try to say something about the population using statistical inference.

Supplementary Information The online version contains supplementary material available at https://doi.org/10.1007/978-3-031-53142-2_3.

B. Karlsen Kivedal, *Applied Statistics and Econometrics*, https://doi.org/10.1007/978-3-031-53142-2_3

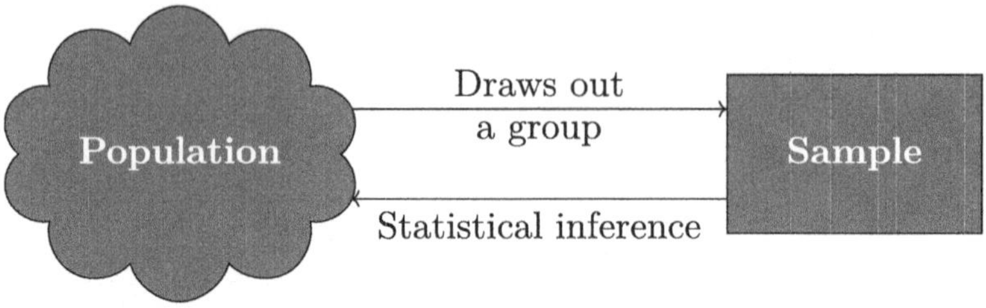

Fig. 3.1 Statistical inference

Note that the population can be seen as an uncertain quantity since one usually does not have information about the entire population. Sometimes, however, we may have data about the entire population, such as if we have data for all relevant companies or anonymous registry data from all tax returns in the country. We will then use statistical inference to say something about a population that is larger than the one we have observations about, such as future cases or other comparable countries (Alexander, 2015).

In this book, we start from the premise that we have a dataset and will perform analyses (statistical inference) based on the given dataset using quantitative methods. The data can come from public sources such as a national statistical agency or the World Bank, or it can be data we have collected from surveys or from companies and organizations. This means that we focus on the use of secondary data sources; that is, data collected by others. For the collection of primary data, there are many challenges and considerations that are also important to focus on, but for this, the reader is referred to other literature, see, e.g., Vannette and Krosnick (2018).

Econometrics and statistics are terms that are often used interchangeably, and both terms are included in the title of this book. While statistics is a field that contains many theories, methods, and tools, econometrics is often referred to as the application of statistics to analyze questions within economics. Slightly different tools are often used in business administration and other fields in social sciences, as well as other sciences. There is some overlap between tools used in econometrics and other statistical tools, and topics that traditionally belong to other fields will also be treated in this book.

3.2 Prerequisites

Some basic tools that are used in this book require a basic understanding of mathematical and statistical concepts. The most important and relevant topics are presented here, focusing on basic notation which is used throughout the book.

3.2.1 Summation

As a summation sign, we typically use the greek letter for big sigma: Σ. If we have three observations of the variable X:

$$X_1 = 3, X_2 = 4 \text{ and } X_3 = 2$$

where the subscripts 1, 2, and 3 indicate the observation number i. Hence, the first observation of X_i, X_1, has a value of 3, the second observation X_2 has the value 4 and the third observation X_3 has the value 2. The sum of these adds to $3 + 4 + 2 = 9$ and we write the sum of these as

$$\sum_{i=1}^{3} X_i = X_1 + X_2 + X_3 = 3 + 4 + 2 = 9$$

We can read the expression including the summation sign $\sum_{i=1}^{3} X_i$ as "the sum of X from observation $i = 1$ to 3". Usually, we see

$$\sum_{i=1}^{n} X_i$$

which is the sum of all values of X from observation number $i = 1$ until the final observation of the sample n (given that the sample size is n, which we usually have).

3.2.2 Derivation

For the function $f(x)$, the derivative at the point where $x = x_1$ is defined as:

$$f'(x_1) = \lim_{\Delta x \to 0} \frac{\Delta y}{\Delta x} = \lim_{\Delta x \to 0} \frac{f(x_1 + \Delta x) - f(x_1)}{\Delta x}$$

where the operator Δ implies change. It is often also written as

$$y' = \frac{dy}{dx} = \frac{df}{dx}$$

Hence, the derivative expresses the change in the dependent variable if the independent variable increases by a small amount. For the function

$$f(x) = y = ax + b$$

where a and b are constants, y is the dependent variable and x is the independent variable, the derivative is

$$\frac{dy}{dx} = a$$

This implies that if x change by a small amount, y change by a. This corresponds to the slope coefficient of a straight line. More about this in Sect. "Derivation Rules". The general rule is that the derivative of $y = ax^c$ (where a and c are constants) is $dy/dx = c \cdot a \cdot x^{a-1}$.

As shown by the tangents in Fig. 3.2, the slope of the polynomial function is 0.7 when $x = 1$ and 1.1 when $x = 3$. Using the general rule of the derivative, we have that the derivative of $y = 2 + 0.5x + 0.1x^2$ is

$$\frac{dy}{dx} = 0.5 + 0.2x$$

Hence, if $x = 1$, the derivative is $0.5 + 0.2 \cdot 1 = 0.7$ and if $x = 3$, the derivative is $0.5 + 0.2 \cdot 3 = 1.1$. This corresponds to the slopes of the tangents shown in the figure. The derivative thus tells us the effect on y if x increase by $\Delta x = 1$. If x increases from 1 to 2, the effect on y is 0.7 and if x increases from 3 to 4, the effect on y is 1.1. This is a non-linear effect, since the effect on y of a change in x depends on the level of x. A linear function such as $y = 3x + 2$ has the derivative $dy/dx = 3$, which implies that the effect on y if x increase by one will always be 3.

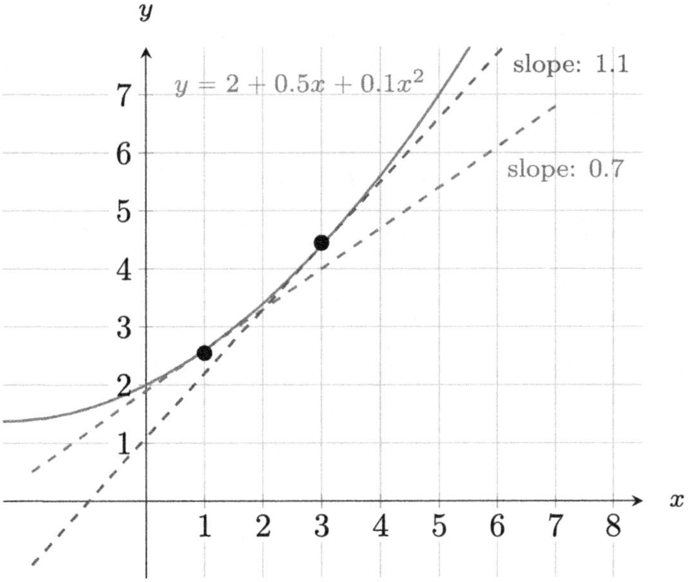

Fig. 3.2 The polynomial function $y = 2 + 0.5x + 0.1x^2$ and its tangents for $x = 1$ and $x = 3$, with slope of 0.7 and 1.1, respectively.

Derivation Rules

The general rule mentioned above (the derivative of $y = ax^c$ is $dy/dx = c \cdot a \cdot x^{a-1}$) results in a number of useful derivation rules:

$$
\begin{array}{ll}
y = b + ax & \frac{dy}{dx} = a \\
y = b + ax^j & \frac{dy}{dx} = jax^{j-1} \\
y = b + a\frac{1}{x} & \frac{dy}{dx} = \frac{-a}{x^2} \\
y = b + ax + cx^2 & \frac{dy}{dx} = a + 2cx \\
y = b + ax = cx^2 + dx^3 & \frac{dy}{dx} = a + 2cx + 3dx^2
\end{array}
$$

When we deal with estimated regression models, as in this book, these can be written as (subscript *i* is omitted to simplify the notation, and we use ∂ instead of *d* as the difference operator which is more general and needed if we have multiple explanatory variables in the model as this implies the partial effect holding other variables constant):

$$
\begin{array}{ll}
\hat{Y} = \hat{\beta}_0 + \hat{\beta}_1 X & \frac{\partial \hat{Y}}{\partial X} = \hat{\beta}_1 \\
\hat{Y} = \hat{\beta}_0 + \hat{\beta}_1 X^j & \frac{\partial \hat{Y}}{\partial X} = j\hat{\beta}_1 X^{j-1} \\
\hat{Y} = \hat{\beta}_0 + \hat{\beta}_1 \frac{1}{X} & \frac{\partial \hat{Y}}{\partial X} = \frac{-\hat{\beta}_1}{X^2} \\
\hat{Y} = \hat{\beta}_0 + \hat{\beta}_1 X + \hat{\beta}_2 X^2 & \frac{\partial \hat{Y}}{\partial X} = \hat{\beta}_1 + 2\hat{\beta}_2 X \\
\hat{Y} = \hat{\beta}_0 + \hat{\beta}_1 X + \hat{\beta}_2 X^2 + \hat{\beta}_3 X^3 & \frac{\partial \hat{Y}}{\partial X} = \hat{\beta}_1 + 2\hat{\beta}_2 X + 3\hat{\beta}_3 X^2
\end{array}
$$

3.2.3 *Expectations*

Within probability theory, we often talk about expectations, and use the expectations operator *E*. The expected value of a variable *X* is thereby written as $E(X)$.

If we consider that a variable *X* which measures the price of a stock has three possible outcomes: $X_1 = 110$ with a probability of 60% ($p_1 = 0.60$), $X_2 = 80$ with a probability of 30% ($p_2 = 0.30$) and $X_3 = 150$ with a probability of 10% ($p_3 = 0.10$), its expected value is

$$E(X) = p_1 X_1 + p_2 X_2 + p_3 X_3 = 0.60 \cdot 110 + 0.30 \cdot 80 + 0.10 \cdot 150 = 105$$

In general, this can be written as

$$E(X) = \sum_{i=1}^{n} p_i \cdot X_i$$

for *n* different outcomes. When rolling a dice with the outcomes 1 to 6 and a probability of $p_i = 1/6$ for each outcome, we thus get

$$E(X) = 1/6 \cdot 1 + 1/6 \cdot 2 + 1/6 \cdot 3 + 1/6 \cdot 4 + 1/6 \cdot 5 + 1/6 \cdot 6 = 3.5$$

Hence, the expected value of a dice throw is 3.5.

3.3 Different Summary Statistics

From a sample, we can, for example, calculate the sample's mean, variance, standard deviation, and confidence intervals. In addition, we can calculate the covariance and correlation between several variables. We can then use these calculations as estimates of the population's mean and standard deviation, etc. The calculations in this chapter involve sample and not population measures.

3.3.1 *Central Tendency and Dispersion*

Measures of Central Tendency

The perhaps most used measure of central tendency is the **mean**.[1] This is also often referred to as the **average**. The mean is calculated as the sum of all values divided by the number of observations. For n observations of the variable X_i for all i units, we then get the sample mean

$$\bar{X} = \frac{\sum_{i=1}^{n} X_i}{n} \tag{3.1}$$

If we use the dataset containing wage and education shown in Table 3.1, we have that the mean of Y is:

$$\begin{aligned}\bar{Y} &= \frac{\sum_{i=1}^{n} Y_i}{n} = \frac{320 + 290 + 170 + 480 + 510 + 410 + 650 + 370}{8} \\ &= \frac{3,200}{8} = 400\end{aligned}$$

and the mean of X is:

$$\bar{X} = \frac{\sum_{i=1}^{n} X_i}{n} = \frac{2+3+1+5+5+4+7+5}{8} = \frac{32}{8} = 4$$

The mean weekly wage for the eight people is therefore 400 EUR, while the mean education after high school is four years.

In addition, we can report the **median** which is the centered value if we align all of the numbers in ascending (or descending) order. For X_i, they are 1, 2, 3, 4, 5, 5, 5, 7, and for Y_i 170, 290, 320, 370, 410, 480, 510, 650.

Since we have eight values, there is no single centered number. The median will therefore be the mean between the two values in the middle. For X_i, the two centered values are $X_6 = 4$ and $X_4 = 5$, such that we get the median $\frac{4+5}{2} = 4.5$. The median for Y_i is found in the same manner, and is $\frac{370+410}{2} = 390$.

The **mode** is also an often used measure of central tendency since it is the number that occurs most frequently. This is often not useful in a small sample. We do not have a mode for Y_t since no persons have the same wage, but for X_i the mode is 5.

[1] We only deal with the arithmetic mean here.

Measures of Dispersion

Since the observations vary around the mean, we often calculate the **standard deviation** to say something about what is a common deviation from the mean. If there are large differences between the wages of the individuals in a sample, the standard deviation will be large, while if the differences are small, the standard deviation will be low. If all the individuals in a sample have the same wage, then everyone will have the mean wage and the standard deviation will be zero. The standard deviation is calculated as the square root of the variance (or, the **variance** is calculated as the square of the standard deviation). The sample standard deviation s is calculated as

$$s = \sqrt{\frac{\sum_{i=1}^{n}(X_i - \bar{X})^2}{n-1}} \tag{3.2}$$

And the variance is

$$s^2 = \frac{\sum_{i=1}^{n}(X_i - \bar{X})^2}{n-1} \tag{3.3}$$

Hence, we may calculate the sample variance as a first step before we calculate the standard deviation.

This implies that we have to calculate the deviation between each observation X_i and the mean $\bar{X}$, and then square these deviations to find the sum of these. The sum of the squared deviations must then be divided by the number of observations n minus 1. This means that we have $n - 1$ **degrees of freedom** when we calculate the standard deviation or variance. Since we have to calculate the mean $\bar{X}$ first, we have already "used one degree of freedom", so we must divide by $n - 1$.

For the 8 people we have observed wages for, we can calculate the standard deviation by finding the sum of the squared deviations:

i	*1*	*2*	*3*	*4*	*5*	*6*	*7*	*8*	*Sum*
X	320	290	170	480	510	410	650	370	3,200
$X - \bar{X}$	−80	−110	−230	80	110	10	250	−30	0
$(X - \bar{X})^2$	6,400	12,100	52,900	6,400	12,100	100	62,500	900	153,400

We then get the sample standard deviation

$$s = \sqrt{\frac{\sum_{i=1}^{n}(X_i - \bar{X})^2}{n-1}} = \sqrt{\frac{153,400}{7}} = \sqrt{21,914.29} = 148.03 \tag{3.4}$$

which is an estimate of the population's standard deviation.

Central Tendency and Dispersion in Gretl

In Gretl, we can choose `View > Summary statistics` to find the mean and standard deviation, as well as other calculations. We often call this descriptive statistics. Here we get the mean, median, and standard deviation, in addition to many other measures. Thus, we get descriptive statistics for the eight observations of wage and education.

```
             Mean        Median       Minimum       Maximum
wage       400.00        390.00        170.00        650.00
edu        4.0000        4.5000        1.0000        7.0000

         Std. Dev.         C.V.      Skewness   Ex. kurtosis
wage       148.03       0.37009       0.16157      -0.59294
edu        1.9272       0.48181      -0.12801      -0.84615

          IQ range   Missing obs.
wage       205.00            0
edu        2.7500            0
```

Central Tendency and Dispersion in R

To find the minimum, maximum, mean, and quartiles in R, we can write `summary()` where the name of the variable/vector we want to find descriptive statistics about is filled in the parentheses. `mean()` gives the mean and `sd()` the standard deviation of the variable. There are also other packages that can be used to easily find different measures. This is shown in Fig. 3.3, where we find descriptive statistics for our data on wage and education.

R code 3.1: Descriptive statistics in R

```
summary(variablename)
mean(variablename)
sd(variablename)
```

We thus get

```
> summary(wage)
Min. 1st Qu.  Median    Mean 3rd Qu.    Max.
170.0   312.5   390.0   400.0   487.5   650.0
> sd(wage)
[1] 148.0347

> summary(edu)
Min. 1st Qu.  Median    Mean 3rd Qu.    Max.
1.00    2.75    4.50    4.00    5.00    7.00
> sd(edu)
[1] 1.927248
```

3.3.2 *Interval Estimate: Confidence Interval*

To say something about the uncertainty in the estimate for the population mean with an explicit probability that provides a more precise measure than just looking at the mean and standard deviation, we can calculate a **confidence interval** for the population mean. If we know the population's standard

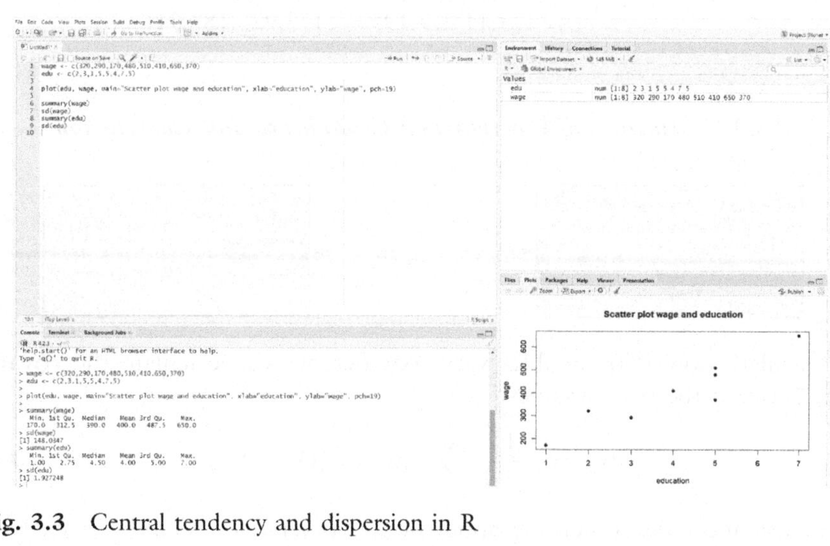

Fig. 3.3 Central tendency and dispersion in R

deviation, we can calculate this as

$$\bar{X} \pm z_{\alpha/2} \cdot \frac{\sigma}{\sqrt{n}} = \left[\bar{X} - z_{\alpha/2} \cdot \frac{\sigma}{\sqrt{n}}, \bar{X} + z_{\alpha/2} \cdot \frac{\sigma}{\sqrt{n}}\right] \tag{3.5}$$

where $z_{\alpha/2}$ is the standard normal distribution's quantile and α is the significance level, e.g., $\alpha = 0.05$. A confidence interval will say something about the uncertainty around the point estimate by the population mean. $(1-\alpha) \cdot 100\%$ of confidence intervals will contain the true population mean at a significance level of α.

If we know the population's standard deviation for the height of 18-year-old men, which we assume is $\sigma = 7$, we have for a significance level of 5

$$\bar{X} \pm z_{\alpha/2} \cdot \frac{\sigma}{\sqrt{n}} = 181.11 \pm 1.96 \cdot \frac{7}{\sqrt{9}} = 181.11 \pm 4.573 = [176.54,\ 185.68]$$

If we do not know the population's standard deviation, we use the t-distribution and the sample standard deviation. We then have

$$\bar{X} \pm t_{\alpha/2,n-1} \cdot \frac{s}{\sqrt{n}} = \left[\bar{X} - t_{\alpha/2,n-1} \cdot \frac{s}{\sqrt{n}}, \bar{X} + t_{\alpha/2,n-1} \cdot \frac{s}{\sqrt{n}}\right] \tag{3.6}$$

where $n-1$ is the number of degrees of freedom. For our example with the height of 9 men we then get:

$$\begin{aligned}\bar{X} \pm t_{\alpha/2,n-1} \cdot \frac{s}{\sqrt{n}} &= \bar{X} \pm t_{0.025,8} \cdot \frac{s}{\sqrt{n}} = 181.11 \pm 2.306 \cdot \frac{9.36}{\sqrt{9}} \\ &= 181.11 \pm 7.195 = [173.92,\ 188.31]\end{aligned}$$

3.3.3 Measures of Covariation: Covariance and Correlation

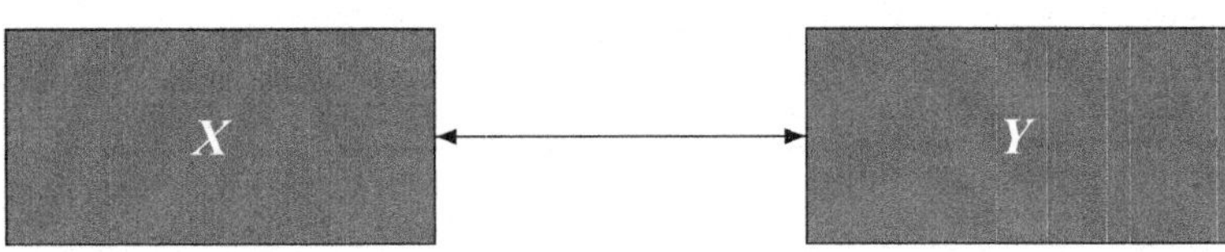

To calculate how two variables vary together, we can calculate the **covariance** between the two variables:

$$s_{XY} = \frac{1}{n-1} \cdot \sum (X_i - \bar{X})(Y_i - \bar{Y}) \tag{3.7}$$

The covariance says something about how two variables X and Y vary with each other. If the covariance is 3, 30, or 300, it is not very informative about how strong the connection is without knowing more about the observations.

By calculating the **correlation** coefficient, we get a standardized measure of whether the covariation is strong or weak. The correlation suggests how strong linear covariation there is between X and Y, and whether it is positive or negative. (Remember that covariation is not the same as correlation.)

- $r = -1$: Perfect negative linear covariation
- $r = 1$: Perfect positive linear covariation
- $r = 0$: No linear covariation

$$r_{XY} = \frac{s_{XY}}{s_X \cdot s_Y} = \frac{\frac{1}{n-1} \cdot \sum(X_i - \bar{X})(Y_i - \bar{Y})}{\sqrt{\frac{1}{n-1} \cdot \sum(X_i - \bar{X})^2} \cdot \sqrt{\frac{1}{n-1} \cdot \sum(Y_i - \bar{Y})^2}}$$

$$= \frac{\sum(X_i - \bar{X})(Y_i - \bar{Y})}{\sqrt{\sum(X_i - \bar{X})^2} \cdot \sqrt{\sum(Y_i - \bar{Y})^2}}$$

$$r_{XY} = \frac{s_{XY}}{s_X \cdot s_Y} = \frac{\sum_{i=1}^{n}(X_i - \bar{X})(Y_i - \bar{Y})}{\sqrt{\sum_{i=1}^{n}(X_i - \bar{X})^2} \cdot \sqrt{\sum_{i=1}^{n}(Y_i - \bar{Y})^2}} \tag{3.8}$$

Example
To calculate the correlation between wage and education from the data shown in Table 3.1 manually (by hand or via spreadsheet without using a specific code) we therefore need, as shown in (3.8), $\sum_{i=1}^{n}(X_i - \bar{X})(Y_i - \bar{Y})$ as well as the square root of $\sum_{i=1}^{n}(X_i - \bar{X})^2$ and $\sum_{i=1}^{n}(Y_i - \bar{Y})^2$. We thus need to calculate the deviation between observed values and the mean value for X and Y, as well as the sum of the product of the deviations and the sum of the squared deviations. We can do this by hand or in a spreadsheet and get:

Table 3.1 Calculation of the correlation coefficient between *X* (education) and *Y* (wage)

X	Y	$X - \bar{X}$	$Y - \bar{Y}$	$(X - \bar{X})(Y - \bar{Y})$	$(X - \bar{X})^2$	$(Y - \bar{Y})^2$
2	320	−2	−80	160	4	6,400
3	290	−1	−110	110	1	12,100
1	170	−3	−230	690	9	52,900
5	480	1	80	80	1	6,400
5	510	1	110	110	1	12,100
4	410	0	10	0	0	100
7	650	3	250	750	9	62,500
5	370	1	−30	−30	1	900
32	3,200	0	0	1,870	26	153,400

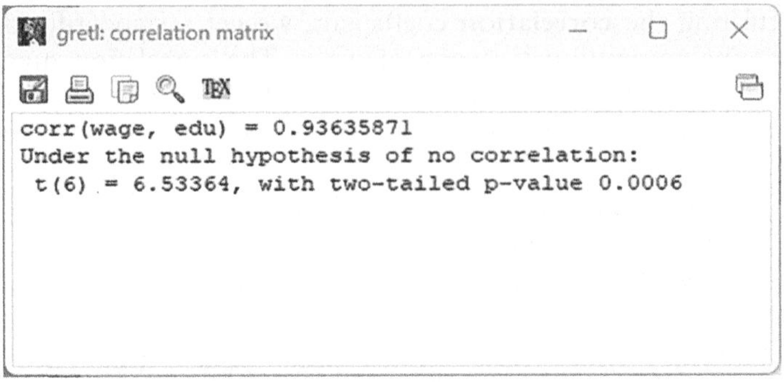

Fig. 3.4 Correlation matrix for two variables

Where we have that $\bar{X} = 4$ and $\bar{Y} = 400$. This gives

$$r_{XY} = \frac{\sum_{i=1}^{n}(X_i - \bar{X})(Y_i - \bar{Y})}{\sqrt{\sum_{i=1}^{n}(X_i - \bar{X})^2} \cdot \sqrt{\sum_{i=1}^{n}(Y_i - \bar{Y})^2}}$$

$$= \frac{1870}{\sqrt{26} \cdot \sqrt{153400}} = \frac{1870}{1997.10} = 0.936$$

Hence, we have a correlation coefficient of 0.936, which tells us that there is a strong positive linear covariance between X (education) and Y (wage).

Correlation in Gretl
To find the correlation between variables, you can go to `View > Correlation matrix`. By selecting education and wage, you will then be given the correlation of 0.936 (Fig. 3.4).

If we have multiple variables, we get a correlation matrix that calculates the correlation between all possible pairs of variables.

We also see that we are given a test statistic and a p-value. This is for a test of whether the correlation is significantly different from zero. More on this in Sect. 4.4.

Correlation in R
To find the correlation between two variables in R, we can use the function `cor()`, where we enter the variables with a comma in between in the parenthesis. For our example, `cor(wage,education)` will give `0.9363587`. We can also use the function `cor.test()` to test if the correlation is significant (more on this in Sect. 4.4). `cor.test(wage,education)` gives

```
Pearson's product-moment correlation

data:  wage and education
t = 6.5336, df = 6, p-value = 0.000614
alternative hypothesis: true correlation is not equal to 0
95 percent confidence interval:
0.6810815 0.9886765
sample estimates:
cor
0.9363587
```

R code 3.2: Correlation in R

```
cor(wage,education)
cor.test(wage,education)
```

3.4 Probability Distributions

Since we want to say something about the population using a sample, we must deal with the fact that there is a certain degree of uncertainty associated with our calculations. In order to say something about this uncertainty, we use various probability distributions. A probability distribution implies that there is a specific relationship between the expected value (population mean) and the dispersion.

3.4.1 Normal Distribution

Variables that are often normally distributed include the height of people, blood pressure, grades, and machine-produced and other mass-produced items, such as the weight of a bag of candy. We can imagine that we know that the population mean for 18-year-old men is 180 cm, while the population standard deviation is 7 cm: $\mu = 180$ cm, and $\sigma = 7$ cm, as shown in Fig. 3.5. The normal distribution is defined such that

- ± 1.96 standard deviations cover 95 % of the population
- ± 2 standard deviations cover 95.44 % of the population
- ± 2.576 standard deviations cover 99 % of the population
- ± 3 standard deviations cover 99.74 % of the population

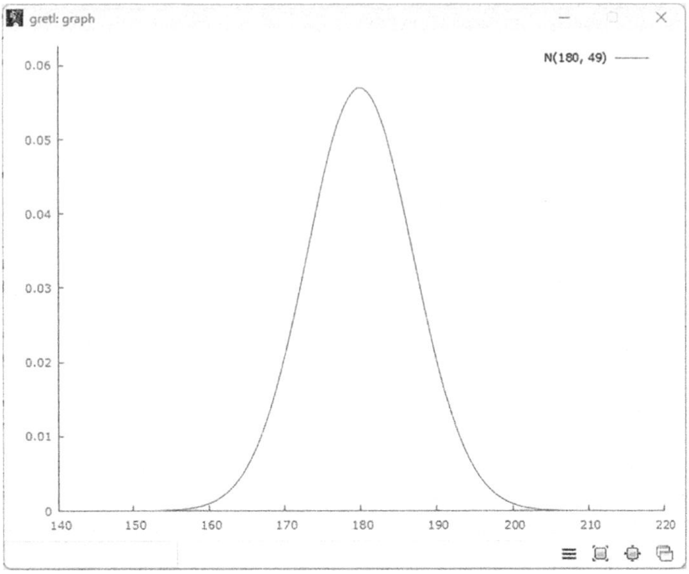

Fig. 3.5 Normal distribution. The height of 18-year-old men

If we go up and down two standard deviations from the population mean, we thus have that there is a 95.44 % probability that a randomly drawn 18-year-old man has a height between 166 and 196 cm.

3.4.2 Standard Normal Distribution

We can rewrite (normalize) the values for a normal distribution so that the population mean becomes 0 and the standard deviation becomes 1. The distribution we then get, we call the **standard normal distribution**, shown in Fig. 3.6. We can do this regardless of what the population mean and the population standard deviation are. In the example above, we had a normal distribution with expectation $\mu = 180$ and standard deviation $\sigma = 7$ (and variance $\sigma^2 = 7^2 = 49$). A standard normal distribution will have expectation $\mu = 0$ and standard deviation $\sigma = 1$ (and variance $\sigma^2 = 1^2$). If X is normally distributed, then $Z = \frac{X-\mu}{\sigma}$ will be standard normally distributed. To transform a normal distribution into a standard normal distribution, we can therefore use the values for X_i and the formula $Z_i = \frac{X_i-\mu}{\sigma}$ to calculate the values of the standard normally distributed variable Z. If we have three observations (three 18-year-old men):

- $X_1 = 180$

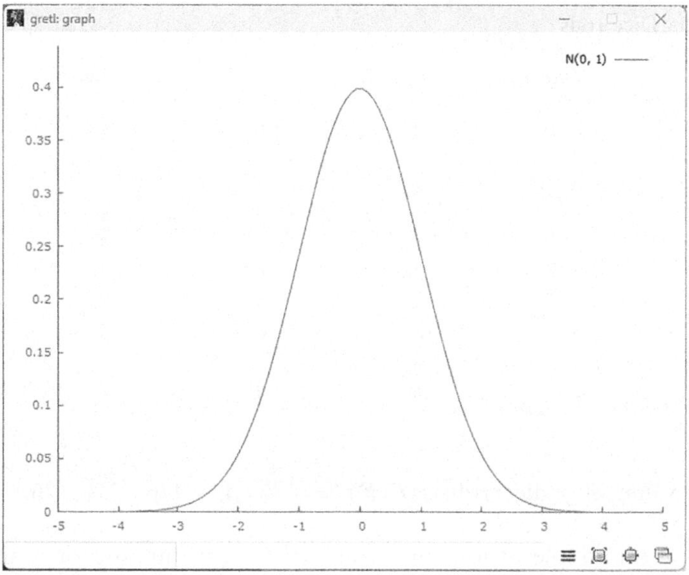

Fig. 3.6 Standard normal distribution

- $X_2 = 194$
- $X_3 = 173$

we thus have that $\frac{X_i - \mu}{\sigma} = \frac{X_i - 180}{7}$ so that

- $Z_1 = \frac{X_1 - \mu}{\sigma} = \frac{180-180}{7} = \frac{0}{7} = 0$
- $Z_2 = \frac{X_2 - \mu}{\sigma} = \frac{194-180}{7} = \frac{14}{7} = 2$
- $Z_3 = \frac{X_3 - \mu}{\sigma} = \frac{173-180}{7} = \frac{-7}{7} = -1$

The values $Z_1 = 0$, $Z_2 = 2$ and $Z_3 = -1$ tell us that person number 2 is 2 standard deviations above the expected value, person 3 is one standard deviation below the expected value, and person number 1 is exactly as tall as the expected value.

3.4.3 Central Limit Theorem

Even though the height of 18-year-old men used in the example above—and many other variables—are normally distributed, not all variables follow this distribution. An example of a variable with a **uniform probability distribution** is the outcome of a dice roll. For a normal dice, there is an equal probability ($1/6 = 0.1667 = 16.67$ %) for each possible outcome. The

Table 3.2 Dice rolls

Sample no.	*Dice 1*	*Dice 2*	*Dice 3*	*Dice 4*	*Mean*
1	2	3	5	6	4.00
2	2	5	4	4	3.75
3	6	2	3	2	3.25
4	4	5	3	1	3.25
5	2	4	5	4	3.75
6	4	2	6	2	3.50
7	1	4	1	3	2.25
8	5	2	6	5	4.50
9	2	5	4	1	3.00
10	6	3	4	2	3.75

expected value of a dice roll is $E(X) = 1/6 \cdot 1 + 1/6 \cdot 2 + 1/6 \cdot 3 + 1/6 \cdot 4 + 1/6 \cdot 5 + 1/6 \cdot 6 = 3.5$.

If we take a sample of four dice rolls ($n = 4$), we might, for example, get two, three, five, and six on the dice these four times. This gives a mean of $\bar{X} = 4$, which in itself doesn't tell us much. If we take more such samples, we get, for example, the outcomes shown in Table 3.2.

By looking closer at the mean values, we see that they seem to be normally distributed. This is due to the **central limit theorem**, which states that sample means will always be normally distributed, regardless of the distribution of the population from which the sample is drawn. Hence, we can often use the normal distribution without having to consider the distribution of the population. The mean of the sample means above becomes $\bar{\bar{X}} = 3.5$ (note the two lines above X, which means that we take the mean of the mean values of X). This is related to the **law of large numbers**, which states that the sample mean from a large sample will get closer to the population mean as the sample size increases. The entire sample has a size of $n = 40$ with a mean of $\bar{X} = 3.5$.

3.4.4 *t-distribution*

If we do not know the population's standard deviation σ (which we usually do not since we do not have data on the entire population), we often use the **t-distribution** instead of the normal distribution. The t-distribution also takes into account the sample size as it depends on the number of degrees of freedom (which in turn is calculated from the sample size), making it especially useful for smaller samples.

As we can see from Fig. 3.7, a t-distribution with few degrees of freedom has larger tails and less probability density around the expected value, 0, than the normal distribution. This means that the interval covering plus minus 2 standard deviations for the height of 18-year-old men will be larger if we have a small sample, as small samples provide greater uncertainty when we want to say

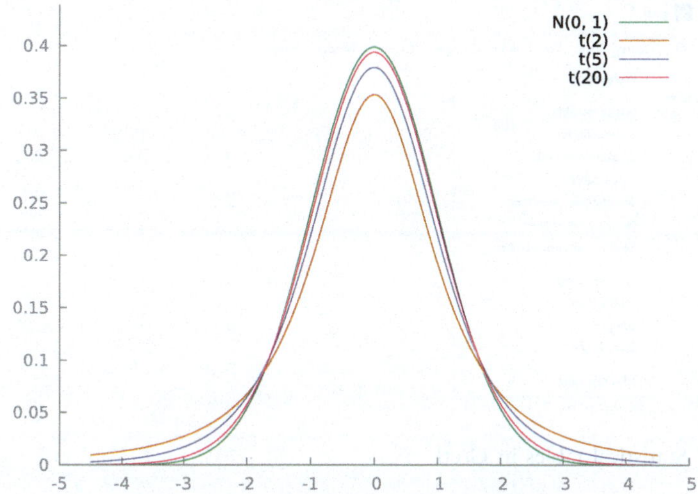

Fig. 3.7 Normal distribution, and the t-distribution with 2, 5, and 20 degrees of freedom

something about the population. At 20 degrees of freedom, the t-distribution is quite close to the normal distribution, and with more degrees of freedom, the t-distribution becomes approximately equal to the normal distribution. The t-distribution is thus most useful if we have few observations.

3.4.5 *Probability Distributions in Gretl*

We often use statistical tables to find the value (quantile) that correspond to a tail probability. In Gretl, you can find this directly by going to `Tools > Statistical Tables`. Here you get the quantiles for the normal distribution, t-distribution, chi-square distribution, F-distribution, binomial distribution, Poisson distribution, Weibull distribution, and Durbin–Watson distribution. By going to `Tools > P-value finder`, you can find the tail probability given a critical value, and through `Tools > Distribution graphs`, you can draw the distributions (Fig. 3.8).

3.4.6 *Probability Distributions in R*

There are functions that can give us values and tail probabilities for different probability distributions in R. For example, we can use `pnorm(z)` to find the tail probability for the value below z in a normal distribution or `qnorm(alpha)` for the value that belongs to the tail probability to the left of alpha. `pnorm(1.96)` will thus give the value 0.975 and `pnorm(−1.96)` give the value 0.025, while `qnorm(0.025)` gives the value −1.96 and `qnorm(0.975)` gives the value 1.96. Alternatively, we

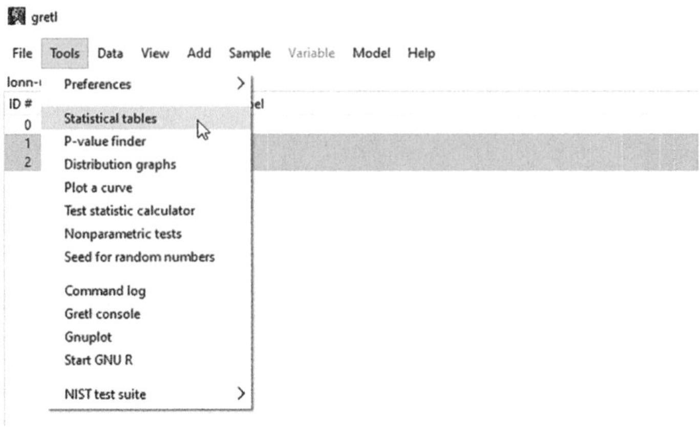

Fig. 3.8 Statistical tables in Gretl

can add `lower.tail=FALSE` to get the value associated with the right-tail probability: `qnorm(0.025, lower.tail=FALSE)` gives the value 1.96.

For the t-distribution, the corresponding `pt(t,df)` is used, where t indicates the quantile and df the number of degrees of freedom, and `qt(alpha,df)`. For the F-distribution and chi-square distribution, we can use `pf(F,df1,df2, lower.tail=FALSE)` and `pchisq(Q,df, lower.tail=FALSE)`, respectively, where F and Q are the quantiles to get the right-tail probability with df1 degrees of freedom in the numerator and df2 in the denominator when using the F-distribution. We find the quantile at a right-tail probability equal to alpha for the F-distribution and chi-square distribution by `qf(alpha,df1,df2,lower.tail=FALSE)` and `qchisq(alpha,df,lower.tail= FALSE)`, respectively.

R code 3.3: Probability distributions in R

Normal distribution (left tail probability):

```
pnorm(-z)
qnorm(alpha)
```

t-distribution (left tail probability):

```
pt(-t,df)
qt(alpha,df)
```

Chi-square distribution (right tail probability):

```
pchisq(Q,df, lower.tail=FALSE)
qchisq(alpha,df, lower.tail=FALSE)
```

F-distribution (right tail probability):

```
pf(F,df1,df2, lower.tail=FALSE)
qf(alpha,df1,df2, lower.tail=FALSE)
```

With values inserted:

```
pnorm(-1.96)
qnorm(0.025)
pt(-2.31,8)
qt(0.025,8)
pf(5.786,2,5,lower.tail=FALSE)
qf(0.05,2,5,lower.tail=FALSE)
pchisq(3.841,1, lower.tail=FALSE)
qchisq(0.05,1, lower.tail=FALSE)
```

This gives

```
> pnorm(-1.96)
[1] 0.0249979
> qnorm(0.025)
[1] -1.959964

> pt(-2.31,8)
[1] 0.02484454
> qt(0.025,8)
[1] -2.306004

> pf(5.786,2,5,lower.tail=FALSE)
[1] 0.05000204
> qf(0.05,2,5,lower.tail=FALSE)
[1] 5.786135

> pchisq(3.841,1, lower.tail=FALSE)
[1] 0.05001368
> qchisq(0.05,1, lower.tail=FALSE)
[1] 3.841459
```

References

Alexander, N. (2015). What's more general than a whole population? *Emerging Themes in Epidemiology, 12*, 1–5.

Vannette, D. L., & Krosnick, J. A. (Eds.). (2018). *The Palgrave handbook of survey research*. Springer International.

CHAPTER 4

Hypothesis Testing

Hypothesis testing of mean values is the core topic of this chapter. The next chapters also involve hypothesis testing, but are not necessarily related to mean values. Hence, it is important to be comfortable with hypothesis testing in general, and this chapter serves as an introduction to hypothesis testing, focusing on formulating hypotheses, as well as the distinction between test statistics, critical values, and *p*-values.

Even though we found that the mean height of the nine 18-year-old men in our sample, which was 181.11 cm, was higher than the expected value of 180 cm, we cannot conclusively state that men are taller than expected (e.g., taller now than they were previously). We must also consider the uncertainty associated with having a small sample and the fact that there is some variability (the standard deviation is not zero) in our sample. To test whether the mean value we found is high enough to conclude that 18-year-old men are taller now than before, we can conduct a hypothesis test.

First, we set up the hypotheses, where we need to formulate the null hypothesis and the alternative hypothesis. Based on the observations and especially the sample mean, standard deviation, and sample size, we can then calculate a test statistic. The test statistic is a value that we compare to a critical value obtained from a probability distribution, and the critical value sets the boundary for a **rejection region**. Whether the test statistic is higher or lower than the critical value, and whether it falls within the rejection region, determines whether we choose to reject or keep the null hypothesis.

Supplementary Information The online version contains supplementary material available at https://doi.org/10.1007/978-3-031-53142-2_4.

B. Karlsen Kivedal, *Applied Statistics and Econometrics*,
https://doi.org/10.1007/978-3-031-53142-2_4

4.1 Formulating Hypotheses

Usually, we formulate the alternative hypothesis based on what our suspicions are, e.g., based on our observations but contrary to established truth. In the case of the nine 18 year olds' height, we would suspect that the population height is greater than 180 cm, so we would formulate the hypotheses as follows:

$$H_0\colon \mu = 180$$
$$H_A\colon \mu > 180$$

We can also set up the hypotheses to be mutually exclusive, even though this is often not done when it is clear that values less than 180 cm are also included in the null hypothesis. In other words, we could write $H_0\colon \mu \leq 180$. In addition, we can have hypothesis tests where the alternative hypothesis states that the population value is less than a certain value or that it is different from a value.

$$H_0\colon \mu = 180$$
$$H_A\colon \mu < 180$$

or

$$H_0\colon \mu = 180$$
$$H_A\colon \mu \neq 180$$

The latter is an example of a two-tailed hypothesis test, while the first two are one-tailed hypothesis tests.

4.2 Test Statistic and Critical Value(s)

The test statistic is calculated based on the sample mean, standard deviation, and sample size, in addition to the value specified in the null hypothesis.

If we know the population standard deviation, we can calculate the test statistic as follows:

$$Z = \frac{\bar{X} - \mu_0}{\sigma/\sqrt{n}}$$

where μ_0 is the mean under the null hypothesis. If we do not know the population standard deviation, we can calculate the test statistic as:

$$T = \frac{\bar{X} - \mu_0}{s/\sqrt{n}}$$

using the sample standard deviation.

		The reality/population	
		H_0 is true	H_0 er incorrect
Our action	Accept H_A and reject H_0	Type 1 error	Correct
	Keep H_0	Correct	Type 2 error

Fig. 4.1 Type 1 and Type 2 errors

To conduct a hypothesis test, we must first decide how much uncertainty we are willing to tolerate. This is done by selecting a **significance level**. The significance level is the probability of rejecting the null hypothesis even if it is true. The lower the probability we are willing to accept for making this error (Type 1 error) is, the higher the probability of keeping the null hypothesis even if it is false (a Type 2 error) will be. Often, a significance level of 5% is used, denoted as $\alpha = 0.05$ (Fig. 4.1).

Significance Level is the tail probability corresponding to the critical value. The value at which we mark 5% of the area completely to the right under the normal distribution curve gives us the critical value.

We, therefore, want to avoid the erroneous rejection of the null hypothesis (Type 1 error). However, if we place too much emphasis on this by choosing a very low significance level, we increase the chances of making a Type 2 error, which we also want to avoid.

Critical Value is found from the probability distribution and is the value corresponding to the tail probability of the significance level. When performing hypothesis testing of the mean, we typically use the normal distribution if we know the population standard deviation, and the t-distribution with $n - 1$ degrees of freedom if we do not know the population standard deviation. However, to find a critical value, we must first select a significance level, which is the probability of making a Type 1 error. In Fig. 4.2, you can see critical values at a 5% significance level using the normal distribution, where we have the significance level using the right, left, and both tail probabilities.

By choosing a significance level, we thus select the probability of making a Type 1 error. If we reject the null hypothesis because the test statistic falls in the rejection region, there will be a probability equal to the significance level that we have wrongly rejected the null hypothesis. In all conclusions from a hypothesis test, there is a probability of making mistakes and that the results we have found are only due to randomness. However, this will also depend on how large the test statistic is, which affects the p-value of the test (more about this in Sect. 4.5).

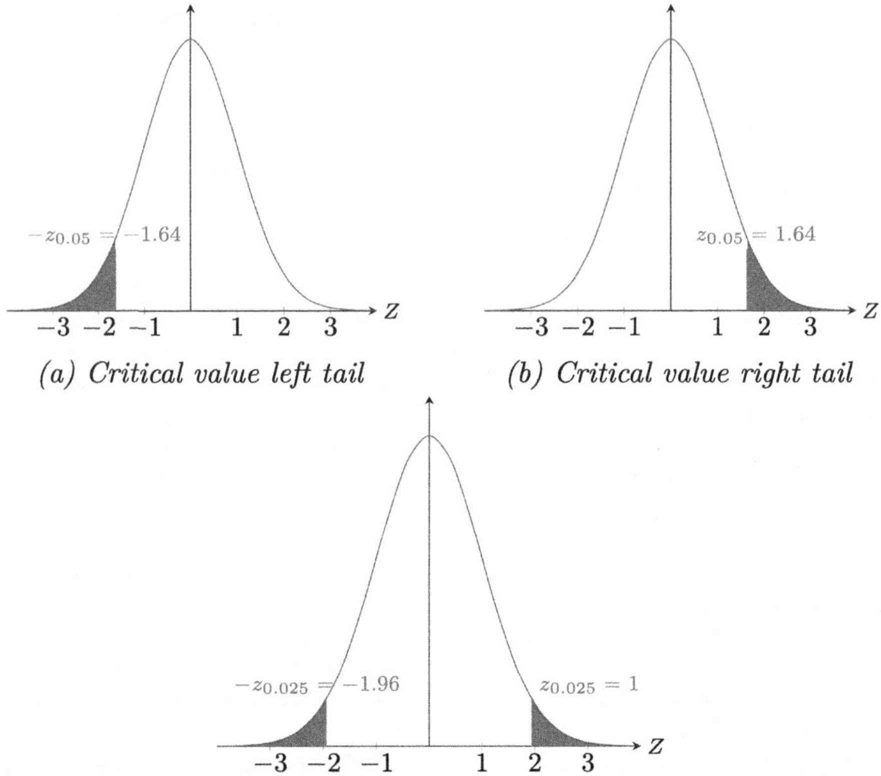

(a) Critical value left tail *(b) Critical value right tail*

(c) Critical values in both tails

Fig. 4.2 The Normal distribution with critical values at a 5% significance level. Red areas show the rejection area

4.2.1 Example: Height of 9 Men—Normal Distribution

We want to test the following hypotheses:

$$H_0 \colon \mu = 180$$
$$H_A \colon \mu > 180$$

where the alternative hypothesis thus suggests that 18-year-old men are taller than 180 cm.

If we know that $\sigma = 7$ for the height of 18-year-old men, we can calculate the test statistic as follows:

$$Z = \frac{\bar{X} - \mu_0}{\sigma/\sqrt{n}} = \frac{181.11 - 180}{7/\sqrt{9}} = \frac{1.11}{2.33} = 0.476$$

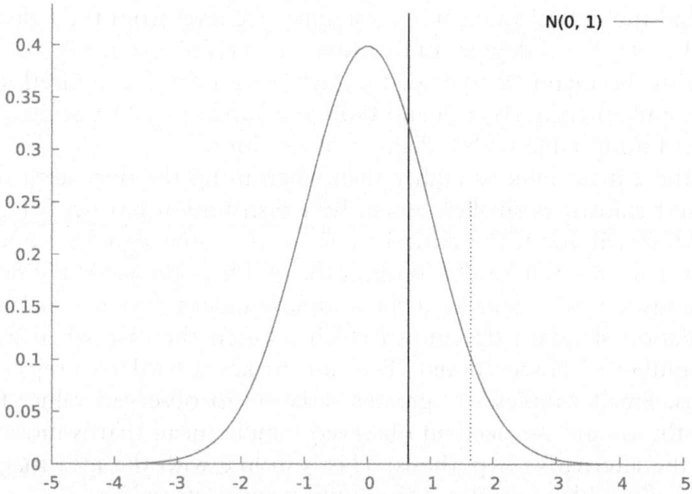

Fig. 4.3 Standard normal distribution with vertical lines representing the test statistic $Z = 0.647$ and the critical value $cv = 1.645$

Choosing a significance level of 5% ($\alpha = 0.05$), we find the critical value in the right tail of the standard normal distribution with a 5% tail area, denoted as $cv = z_\alpha = z_{0.05} = 1.645$. You can find this value using `Tools > Statistical tables` in Gretl or the function `qnorm(alpha)` in R, or in a statistical table for the standard normal distribution.

The rejection region is the area to the right of the critical value. If the test statistic lies in this region, we can reject the null hypothesis. However, the test statistic $Z = 0.647$ in our example does not fall into the rejection region. Therefore, we fail to reject the null hypothesis and cannot conclude that men are taller than 180 cm based on our sample of 9 individuals (see Fig. 4.3).

4.3 Examples in Gretl and R

4.3.1 Height of 9 Men: t*-Distribution*

If we had not known the population standard deviation, σ, we would have to use the sample standard deviation, s, to calculate the test statistic. In addition, we would find the critical value from the t-distribution with $n - 1$ degrees of freedom instead of the normal distribution. We calculated $s = 9.36$ in Eq. (3.4). This gives the test statistic:

$$T = \frac{\bar{X} - \mu_0}{s/\sqrt{n}} = \frac{181.11 - 180}{9.36/\sqrt{9}} = \frac{1.11}{3.12} = 0.356$$

We can find the critical value at a 5% significance level from the t-distribution with $n-1=9-1=8$ degrees of freedom: $cv = t_{0.05}(8) = 1.860$. You can find this value in the menu `Tools > Statistical tables` in Gretl, choosing the `t` tab, and entering `df` as 8 and 0.05 as `right-tail probability`, or in a t-distribution table with 8 degrees of freedom.

Now, the critical value is higher than when using the normal distribution, and the test statistic is smaller. Since the t-distribution has heavier tails than the normal distribution, the critical value (at the same significance level) will be higher. This makes it harder to reject the null hypothesis. At the same time, the test statistic will be smaller if the sample standard deviation is larger than the population standard deviation (which is often the case when we have a limited number of observations). This also makes it harder to reject the null hypothesis. Small samples and greater variation in observed values favor the null hypothesis, and we need an observed sample mean that is more extreme to reject the alternative hypothesis. This is in line with the idea that the null hypothesis should be considered true until proven otherwise.

Hypothesis Testing for Men's Height in Gretl

As shown above, we can find the critical value in Gretl and compare it to the test statistic. But if we have a dataset in Gretl, for example, the height of the 9 individuals mentioned earlier, we can perform the hypothesis test directly in Gretl.

Go to `Tools > Test statistics calculator` and select the variable "`height`". The mean, standard deviation, and sample size (`sample mean`, `std. deviation`, and `sample size`) are then automatically entered. After that, you can fill in the value 180 at the line `H0: mean =` , which is the mean under the null hypothesis, μ_0. By not checking the box for `Assume standard deviation is population value`, we will get Fig. 4.4. If we check this box, we use the normal distribution and the number for `std. deviation` is interpreted as the population standard error. Here, a vertical line is drawn for the test statistic (`test statistic`).

In addition, we will get the following output:

```
Null hypothesis: population mean = 180
Sample size: n = 9
Sample mean = 181.111, std. deviation = 9.36008
Test statistic: t(8) = (181.111 - 180) / 3.12003 = 0.356122
Two-tailed p-value = 0.731
(one-tailed = 0.3655)
```

This output shows the values we entered before the test, along with the test statistic with some calculations. In addition, p-values for both two-tailed and one-tailed tests are provided (see the next subsection). We need to find the critical values from the t-distribution (from a table or Gretl). However, from

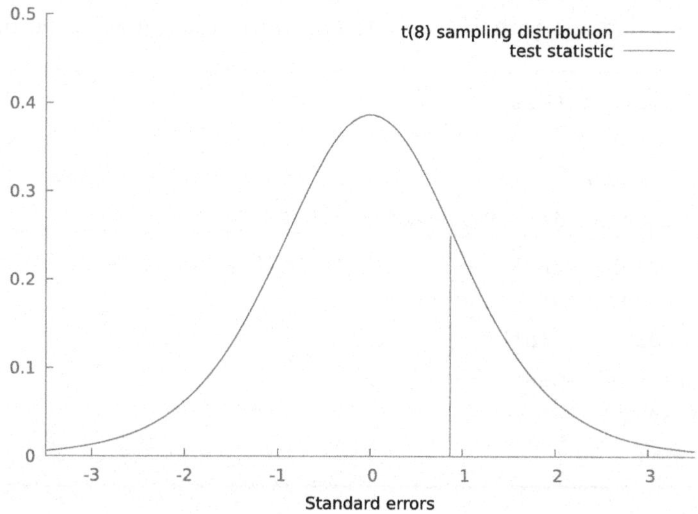

Fig. 4.4 Outcome of hypothesis testing

the p-value (one-tailed test), we can see that it is higher than the significance level, so we conclude that we keep the null hypothesis.

Hypothesis Testing for Men's Height in R

If we enter the data for the height of the 9 men as a variable named `height` in R by typing `height <- c(181, 182, 170, 172, 190, 200, 175, 177, 183)`, we can perform a t-test for the mean using the function `t.test(x, mu = value)`, where `x` is the variable/data vector in R, and `value` is the value specified in the null hypothesis. We can also specify an alternative hypothesis with "greater" as in our example by adding the command `alternative = 'greater'`. For our example, we can write `t.test(height, mu = 180, alternative = 'greater')`. By writing `t.test(height, mu = 180)`, we get a two-tailed test where H_A: $\mu \neq 180$.

R code 4.1: One-Sided Hypothesis Test in R

```
height <- c(181,182,170,172,190,200,175,177,183)
t.test(height,mu=180,alternative = 'greater')
```

The one-sided test with H_A: $\mu > 180$ produces the following output in R:

```
One Sample t-test

data:  hoyde
t = 0.35612, df = 8, p-value = 0.3655
alternative hypothesis: true mean is greater than 180
95 percent confidence interval:
175.3093        Inf
sample estimates:
mean of x
181.1111
```

4.3.2 *Example Batteries: "Less Than" Alternative Hypothesis*

Hypothesis testing is often used for statistical process control, such as examining whether the goods we produce or sell align with our expectations. This can involve the contents of food packaging in terms of quantity or weight, the lifetime of a product, or other properties that are expected to have a certain value. For example, we might suspect that the batteries we sell do not have the same lifespan as stated by the manufacturer.

Assume that we sell batteries. According to the manufacturer, the expected battery lifespan is $\mu = 5000$ hours with a standard deviation of $\sigma = 200$. We suspect a lower lifespan than stated because we have taken a sample of $n = 15$ batteries with an average lifespan of $\bar{X} = 4900$ hours. Is 4900 significantly lower than 5000 hours for us to claim that the batteries do not meet the manufacturer's promises, or is there too much uncertainty associated with our observations to conclude that the manufacturer falls short of expectations? To determine this, we can perform the following hypothesis test:

$$H_0: \mu = 5000$$
$$H_A: \mu < 5000$$

The test statistic then becomes

$$Z = \frac{\bar{X} - \mu_0}{\sigma/\sqrt{n}} = \frac{4900 - 5000}{200/\sqrt{15}} = \frac{-100}{51.640} = -1.936$$

We find the critical value in the normal distribution. At a 5% significance level, we get $cv = -z_\alpha = -z_{0.05} = -1.645$.

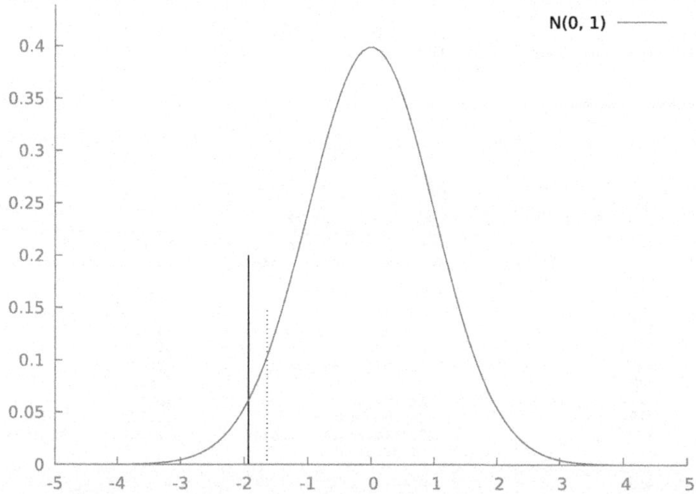

Fig. 4.5 Standard normal distribution with a critical value of $cv = -1.645$ and a test statistic of $Z = -1.936$

Alternating hypothesis with the "less than" sign

If the alternative hypothesis includes the statement "less than" (which can be formulated as $<$), we must use a negative critical value. This is because we are then using the left tail of the probability distribution.

Test statistic and critical value for the example are shown in the standard normal distribution in Fig. 4.5. Here, we can see that the test statistic is in the rejection region. Therefore, we reject the null hypothesis and conclude that the batteries have a shorter lifespan than stated.

Battery Example with Unknown Population Standard Deviation

If we do not know the population standard deviation, we have to use the sample standard deviation. Additionally, in this case, it's advantageous to use the t-distribution to find the critical value, especially when dealing with small samples (where $n < 30$), as we have here. As shown in Fig. 4.6, we have a sample mean of $\bar{X} = 4900$ and a sample standard deviation of $s = 368.39$. These values can also be manually calculated as shown in Table 4.1, where we compute $s = \sqrt{\frac{\sum(X-\bar{X})^2}{n-1}} = \sqrt{\frac{1900000}{14}} = \sqrt{135714.286} = 368.39$.

This gives us the test statistic:

$$T = \frac{\bar{X} - \mu_0}{s/\sqrt{n}} = \frac{4900 - 5000}{368.39/\sqrt{15}} = \frac{-100}{95.118} = -1.05$$

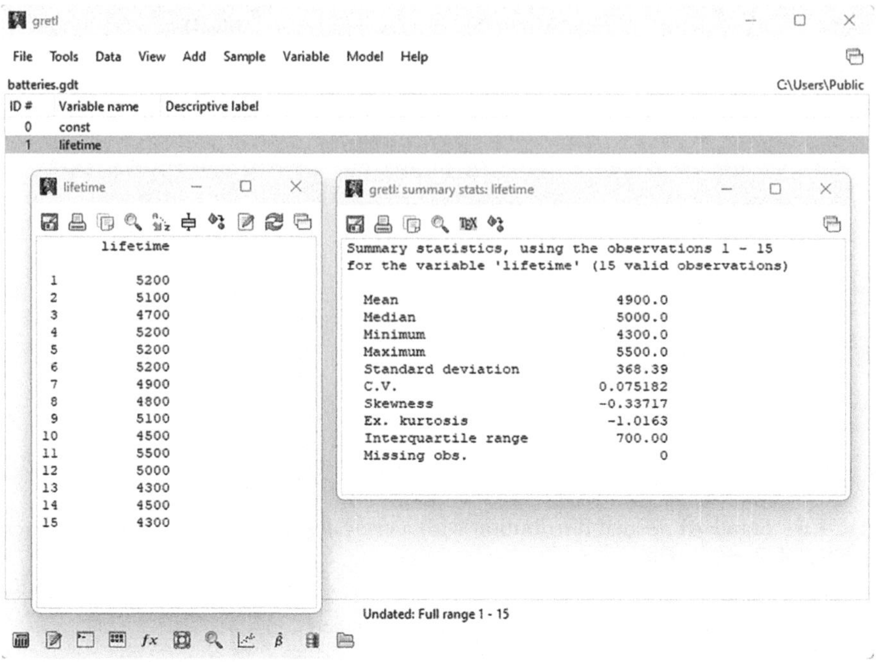

Fig. 4.6 Descriptive statistics of battery lifetimes

Table 4.1 Battery lifetimes

i	X	$X - \bar{X}$	$(X - \bar{X})^2$
1	5,200	300	90,000
2	5,100	200	40,000
3	4,700	−200	40,000
4	5,200	300	90,000
5	5,200	300	90,000
6	5,200	300	90,000
7	4,900	0	0
8	4,800	−100	10,000
9	5,100	200	40,000
10	4,500	−400	160,000
11	5,500	600	360,000
12	5,000	100	10,000
13	4,300	−600	360,000
14	4,500	−400	160,000
15	4,300	−600	360,000
Sum	73,500	0	1,900,000
Mean	4,900		

The critical value at a 5% significance level is found in the t-distribution with $n - 1 = 14$ degrees of freedom, and we need to consider the left tail of the distribution: $cv = -t_{\alpha}(n - 1) = -t_{0.05}(14) = -1.761$. Since the absolute value of the test statistic is less than the absolute value of the critical value, we keep the null hypothesis. We cannot conclude that the batteries have less lifetime than stated.

There are two main reasons why we keep the null hypothesis when using the sample standard deviation and the t-distribution compared to when we used the population standard deviation and the normal distribution. Firstly, we are using the sample standard deviation, $s = 368.39$, which is much larger than $\sigma = 200$. A higher standard deviation causes the test statistic to be lower, moving it closer to the center of the probability distribution, which decreases the probability of it falling into the rejection region. Another important reason is that we are using the t-distribution. The t-distribution has heavier tails than the normal distribution, which moves the rejection region farther away from zero. This can be observed in the critical value of -1.761 when using the t-distribution with 14 degrees of freedom, compared to -1.645 with the standard normal distribution.

Fewer degrees of freedom would further intensify this effect. Smaller sample sizes result in fewer degrees of freedom, leading to heavier tails in the t-distribution. Additionally, smaller samples result in a larger denominator in the test statistic formula. This reduces the absolute value of the test statistic, and hence, reduces the probability of rejecting the null hypothesis. It is therefore harder to reject the null hypothesis with smaller sample sizes both due to effect on the test statistic and the critical value.

Hypothesis Testing, Batteries, in Gretl

To perform the hypothesis tests for batteries, both with a known and unknown population standard deviation, you can use Gretl. Start by navigating to `Tools > Test statistic calculator`. Check the box labeled `Use variable from dataset` and select the variable `lifetime`, which will automatically provide the sample mean, standard deviation, and sample size. Next, enter the value under the null hypothesis, `H0: mean =`, which in this case is 5000. If you check the box `Assume standard deviation is population value`, you should enter $\sigma = 200$ as `std.deviation`. If not, you can use the t-distribution and $s = 368.39$. The results from Gretl are displayed in Figs. 4.7 and 4.8.

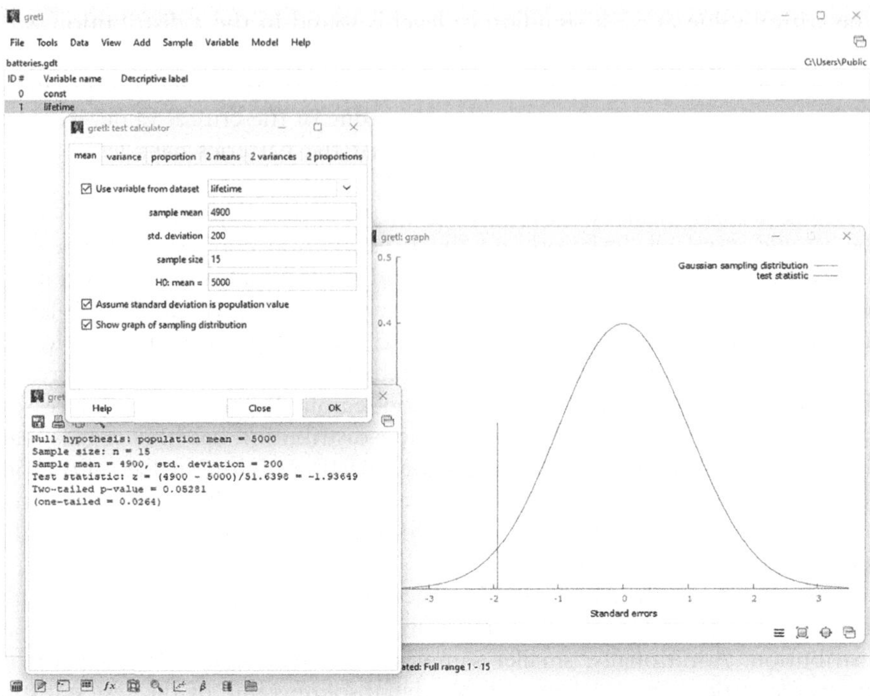

Fig. 4.7 Battery Hypothesis Test in Gretl, Known Population Standard Deviation

Hypothesis Testing, Batteries, in R

R code 4.2: One sided test i R

```
lifetime <- c(5200,5100,4700,5200,5200,5200,4900,4800,5100,
              4500,5500,5000,4300,4500,4300)
t.test(lifetime,mu=5000,alternative = 'less')
```

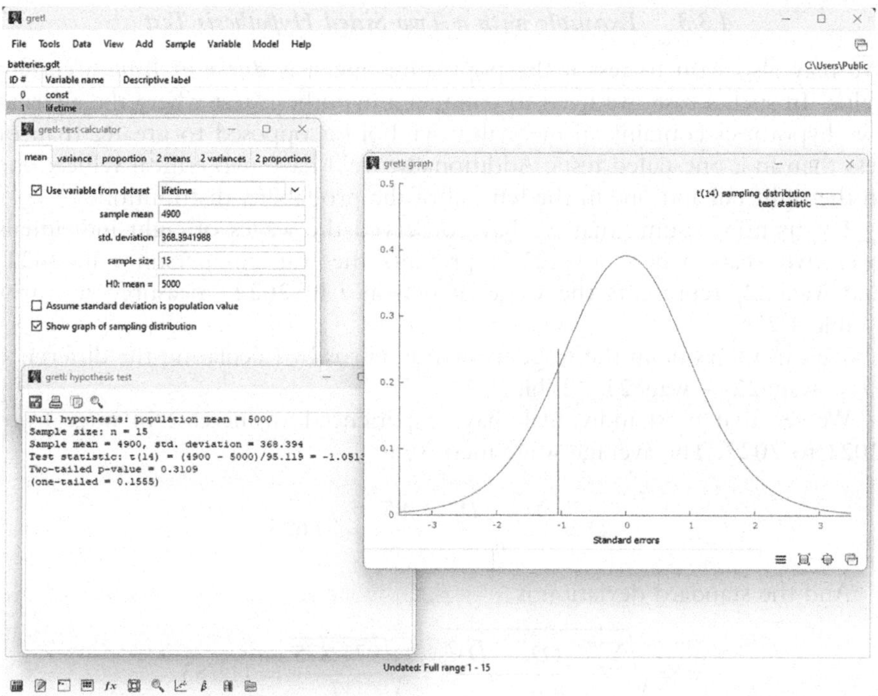

Fig. 4.8 Battery Hypothesis Test in Gretl, Unknown Population Standard Deviation

This yields

```
One Sample t-test

data:  levetid
t = -1.0513, df = 14, p-value = 0.1555
alternative hypothesis: true mean is less than 5000
95 percent confidence interval:
-Inf 5067.534
sample estimates:
mean of x
4900
```

4.3.3 *Example with a Two-Sided Hypothesis Test*

We may also wish to test if the population mean is *different from* a certain value. In such a case, we need to conduct a hypothesis test where the alternative hypothesis contains an inequality symbol (as opposed to greater than or less than in a one-tailed test). Additionally, we'll have two critical values: one in the right tail and one in the left tail of the probability distribution.

Let us now assume that we have observed the wages of eight individuals over two years, where wage21_i represents the wage of person *i* in 2021, and wage22_i represents the wage of person *i* in 2022, measured in euros (Table 4.2).

We can then obtain the wage growth in euros by calculating the difference $D_i = \text{wage22}_i - \text{wage21}_i$ (Table 4.3).

We see that most individuals have experienced an increase in wage from 2021 to 2022. The average wage increase is

$$\bar{D} = \frac{\sum_{i=1}^{8} D_i}{n} = \frac{85}{8} = 10.625$$

And the standard deviation is

$$s = \sqrt{\frac{\sum_{i=1}^{n}(D_i - \bar{D})^2}{n-1}} = \sqrt{\frac{971.875}{7}} = 11.783$$

To test if this is significantly different from zero (meaning we are not concerned with whether there is a wage increase or decrease, but only want to test if the wages are different in 2022 and 2021), we can use the following hypothesis:

$$H_0\colon \mu_D = 0$$
$$H_A\colon \mu_D \neq 0$$

where μ_D is the population mean of wage differences.

Table 4.2 Wages in 2021 (wage21) and 2022 (wage22)

i	*1*	*2*	*3*	*4*	*5*	*6*	*7*	*8*
wage21	315	295	165	470	490	380	630	370
wage22	320	290	170	480	510	410	650	370

Table 4.3 Wage growth in euros

i	*1*	*2*	*3*	*4*	*5*	*6*	*7*	*8*
D	5		−5 5	10	20	30	20	0

We then calculate the test statistic as follows:

$$T = \frac{\bar{D} - \mu_0}{s_D/\sqrt{n}} = \frac{10.625}{11.783/\sqrt{7}} = \frac{10.625}{4.454} = 2.386$$

And find the critical values at a 5% significance level:

$$\pm t_{\alpha/2}(n-1) = \pm t_{0.025}(7) = \pm 2.365$$

We see that the test statistic, in absolute value, is slightly higher than the critical value in absolute value, so we reject the null hypothesis. We hence find that wages are different in 2021 and 2022.

When testing for a difference between two variables where there are observation pairs (e.g., the variable for 2022 includes the same individuals as in 2021, so we can measure the difference for each individual) and performing a two-tailed hypothesis test like in this case, we have conducted what is often known as a **paired *t*-test**.

Two-Tailed Hypothesis Test in Gretl
To conduct the hypothesis test for the difference, go to `Tools > Test statistic calculator` in Gretl. Check the option `Use variable from dataset` and select the variable `D`, after calculating this. Additionally, enter the value 0 in the field `H0: mean =` and click OK. You will then obtain the following output:

```
Null hypothesis: population mean = 0
Sample size: n = 8
Sample mean = 10.625, std. deviation = 11.783
Test statistic: t(7) = (10.625 - 0)/4.16592 = 2.55046
Two-tailed p-value = 0.03808
(one-tailed = 0.01904)
```

This output matches what we obtained when performing the test manually (with slight differences due to rounding).

Two-Tailed Hypothesis Test in R
You can use the `t.test` function for a two-tailed hypothesis test in R, but you need to select `alternative = 'two.sided'`.

R code 4.3: Two-Tailed Hypothesis Test in R

```
t.test(D, mu = 0, alternative = 'two.sided')
```

You will then obtain the following output in R:

```
> t.test(D, mu = 0, alternative = 'two.sided')

One Sample t-test

data:  D
t = 2.5505, df = 7, p-value = 0.03808
alternative hypothesis: true mean is not equal to 0
95 percent confidence interval:
0.7741585 20.4758415
sample estimates:
mean of x
10.625
```

We can also conduct this test directly on the two variables by using the option `paired = TRUE`. This then performs the paired t-test directly.

R code 4.4: Paired t-test

```
t.test(wage22,wage21, var.equal=TRUE,paired = TRUE)
```

This gives

```
> t.test(wage22,wage21, var.equal=TRUE,paired = TRUE)

Paired t-test

data:  wage22 and wage21
t = 2.5505, df = 7, p-value = 0.03808
alternative hypothesis:
true difference in means is not equal to 0
95 percent confidence interval:
0.7741585 20.4758415
sample estimates:
mean of the differences
10.625
```

which is the same result as the two-sided hypothesis test above.

Hypothesis testing, step by step

In summary, the following steps can be followed when conducting a statistical hypothesis test (the order can be flexible):

1. Set up the alternative hypothesis.
2. Set up the null hypothesis.
3. Choose the significance level.
4. Collect data (sample).
5. Calculate the test statistic from the sample.
6. Find the critical value(s) to determine the rejection region.
7. Draw a conclusion:
    - Reject the null hypothesis and accept the alternative hypothesis, *or*
    - Retain the null hypothesis.

4.4 Hypothesis Testing for Correlation

Whether we find a positive or negative, weak or strong correlation, as discussed in Sect. 3.3.3, we may be interested in whether the correlation is statistically significant. This can be tested using a hypothesis test.

$$H_0\colon \rho = 0$$
$$H_A\colon \rho \neq 0$$

Where ρ is the population correlation. We can also use a one-sided test to determine whether there is a positive or negative correlation by setting $\rho > 0$ or $\rho < 0$ as the alternative hypothesis. The test statistic is calculated as

$$T = \frac{r\sqrt{n-2}}{\sqrt{1-r^2}} \tag{4.1}$$

where r is the correlation coefficient for the sample and n is the number of observations. Critical values are found in the t-distribution with $n-2$ degrees of freedom.

To test whether the correlation between education and income ($r_{XY} = 0.936$) in our example with eight individuals is significant, we can calculate the test statistic as follows:

$$T = \frac{r_{XY}\sqrt{n-2}}{\sqrt{1-r_{XY}^2}} = \frac{0.936\sqrt{8-2}}{\sqrt{1-0.936^2}} = \frac{0.936\cdot\sqrt{6}}{\sqrt{0.1239}} = \frac{0.936\cdot 2.4495}{0.352} = 6.513$$

As shown in Sect. 3.3.3, we obtained a test statistic of 6.53 when calculating the correlation in Gretl or R (which is the same as the one calculated here due to rounding).

Critical values are found in the t-distribution with $n-2$ degrees of freedom. For a 5% significance level, the critical values are

$$\pm t_{\alpha/2}(n-2) = \pm t_{0.025}(6) = \pm 2.447$$

We can see that the test statistic falls within the rejection region, allowing us to conclude that there is a significant correlation between income and education.

4.5 *p*-Value

The p-value is a probability value that corresponds to the test statistic. It expresses the probability of observing something as extreme as what we have observed, assuming that the null hypothesis is true. A low p-value, therefore, implies that we can reject the null hypothesis because it is unlikely that our extreme observation is random. On the other hand, a high p-value means that we keep the null hypothesis since there is a high probability that our observation is not a random deviation from the truth. When we choose a significance

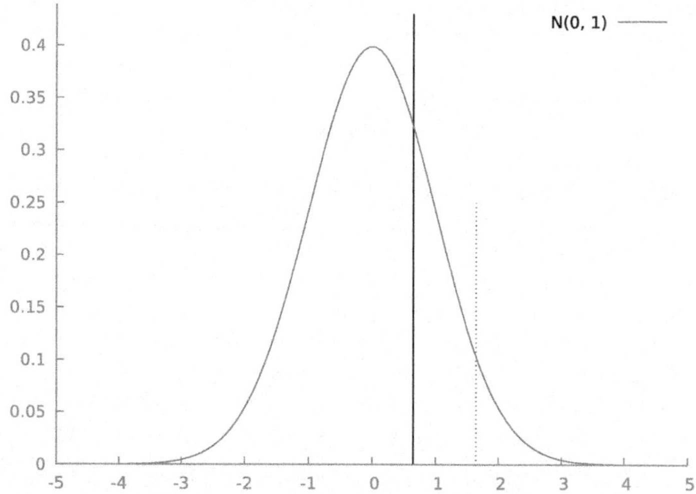

Fig. 4.9 The standard normal distribution with vertical lines for the test statistic $Z = 0.647$ and the critical value $cv = 1.645$

level, we set a threshold for rejecting the null hypothesis if the *p*-value is less than the significance level.

As shown in Fig. 4.3 (replicated below as Fig. 4.9), we had a significance level of 5%. This corresponded to a critical value of 1.645, meaning that the area under the curve to the right of 1.645 (indicated by the vertical line) is 5% of the total area under the curve. The *p*-value is the area under the curve to the right of the test statistic $Z = 0.647$. From the standard normal distribution table, we find that the area under the curve to the right of 0.647 is 0.259 (`Tools > p-value finder` in Gretl). This means that the *p*-value is 0.259, and we keep the null hypothesis since it is not lower than the significance level.

Hypothesis testing using a p-value

If the p-value is *less than* the significance level, we *reject* the null hypothesis.

If the p-value is *greater than or equal to* the significance level, we *keep* the null hypothesis.

CHAPTER 5

Simple Linear Regression

This chapter covers an introduction to simple linear regression. This extends the concept of correlation from Chapter 3, and involves estimating the relationship between two variables. We also cover the main concepts for the ordinary least squares method in this chapter, which is the core for the next chapters. Hypothesis testing of regression coefficients is also covered, and the focus is on relating the output from Gretl and R to what we can obtain manually when working with a small dataset.

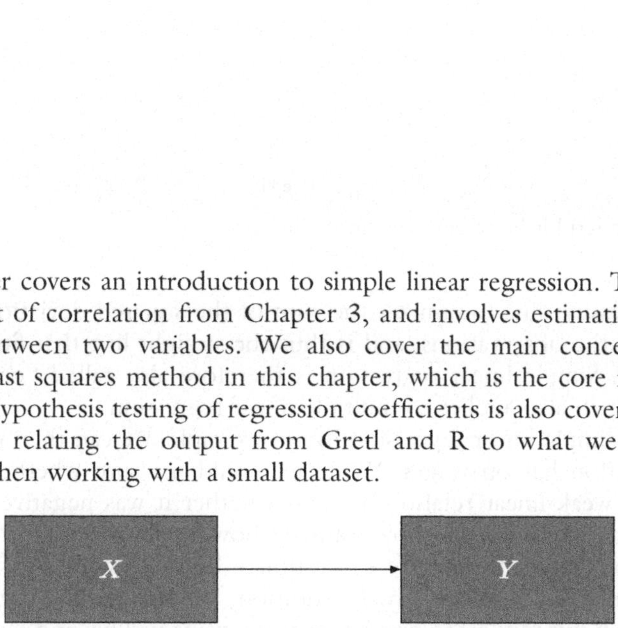

Simple regression implies that we try to estimate a relationship between two variables. This is a linear relationship and is mainly linked to correlation as shown in Sect. 3.3.3.

While correlation measured how strong the linear covariation between two variables was, simple regression will attempt to measure how two variables affect each other. As shown in the scatterplot in Fig. 2.9 (reproduced as Fig. 5.1), we have a line that lies along the points. This is the **regression line**.

Supplementary Information The online version contains supplementary material available at https://doi.org/10.1007/978-3-031-53142-2_5.

B. Karlsen Kivedal, *Applied Statistics and Econometrics*,
https://doi.org/10.1007/978-3-031-53142-2_5

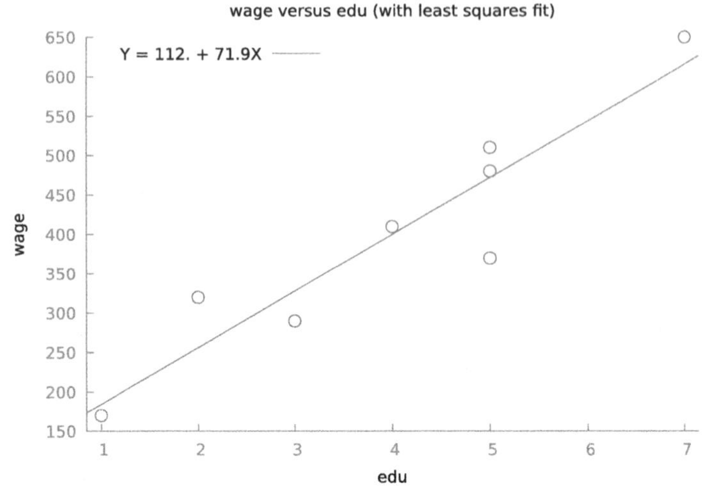

Fig. 5.1 Scatterplot between wage and education

The regression line minimizes the sum of the squared deviations between the line and the observations, and is thus the straight line that fits the observations best. Simple linear regression is therefore also called "ordinary least squares", which also applies to multiple linear regression.

By using simple linear regression in our example, we can then measure the effect education has on wages. While we could measure whether there was a strong or weak linear relationship and whether it was negative or positive through correlation, we can now estimate how a change in education affects wages. Among other things, we can estimate how the wage is affected if we have an extra year of education. In addition, we can use the regression line (which is then our model) to predict what the wage will be, and say something about whether the estimated effects are statistically significant.

5.1 The Regression Line

5.1.1 A Straight Line: Math

From introductory mathematics, we have that the formula for a straight line is

$$y = ax + b \tag{5.1}$$

or

$$y = b + ax$$

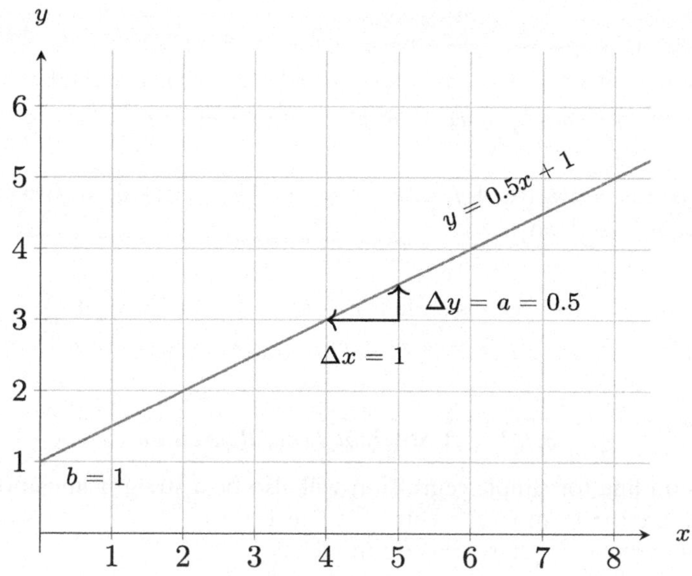

Fig. 5.2 A straight line with a slope coefficient of 0.5 and intersection of 1.0 ($y = 0.5x + 1$)

where a is the **slope coefficient** of the line (how much y increases if x increases by one unit) and b is the **intersection point** of the line (the value of y where the line intersects the y-axis).

Slope coefficient and intersection of a straight line

The slope coefficient, a, of the line shows how much y increases by (or decreases by if $a < 0$) if x increases by 1.

b is the intersection point and shows the value of y if $x = 0$, i.e. the intersection point between the line and the y-axis.

If $a = 0.5$ and $b = 1$, we get $y = 0.5x + 1$ as drawn in Fig. 5.2. Here we see that the line intersects with the y-axis where $y = 1$ (and $x = 0$). If x increases by 1, $\Delta x = 1$, y increases by 0.5 (slope coefficient $a = 0.5$), $\Delta y = 0.5$. The derivative of y with respect to x is $dy/dx = 0.5$, which also shows the effect on y if x increases by 1. See the box below.

Differentiation

The derivative of $y = ax^b$ where a and b are constants and x is a variable becomes $\frac{dy}{dx} = ax^{b-1}$. If $b = 1$, we get $\frac{dy}{dx} = a$.

Thus, the derivative of $y = 2x + 3$ will be $\frac{dy}{dx} = 2$ (the derivative of a constant will be 0).

For an estimated regression line $\hat{Y} = \hat{\beta}_0 + \hat{\beta}_1 X$ we then get $\frac{d\hat{Y}}{dX} = \hat{\beta}_1$

5.1.2 *A Straight Line: Regression*

A regression line for simple regression will also be a straight line and thus be expressed like the formula for a straight line $y = ax + b$ or $y = b + ax$ shown above. We write a simple regression model as

$$\underbrace{Y_i = \beta_0 + \beta_1 X_i}_{\text{straight line}} + u_i \tag{5.2}$$

where we have a straight line ($Y_i = \beta_0 + \beta_1 X_i$) with the variables Y and X and the coefficients β_0 and β_1, in addition to having the **error term** u_i. β_0 is thus the point of intersection of the regression line and β_1 is the slope coefficient.

Y_i and X_i are the actual **observed** values of the variables Y and X, while the regression model will give an **estimated** value of the dependent variable which we write as $\hat{Y}$ ("Y hat"). The deviation between observed and estimated Y gives us the error term $\hat{u}$.

$$\hat{u}_i = Y_i - \hat{Y}_i \tag{5.3}$$

where the objective of the regression model is to minimize (the squared sum of) these deviations. Note that we also put ^ (a "hat") above the error term, since this is also an estimated variable. The error term (often called **residual** or **deviation**) will thus contain everything that is not explained by the straight line. The regression line is then our **model** for how X affects Y.

Hence, we have that estimated Y becomes

$$\hat{Y}_i = \hat{\beta}_0 + \hat{\beta}_1 X_i \tag{5.4}$$

where

$$\hat{\beta}_1 = \frac{\sum(Y_i - \bar{Y})(X_i - \bar{X})}{\sum(X_i - \bar{X})^2} \tag{5.5}$$

$$\hat{\beta}_0 = \bar{Y} - \hat{\beta}_1 \bar{X} \tag{5.6}$$

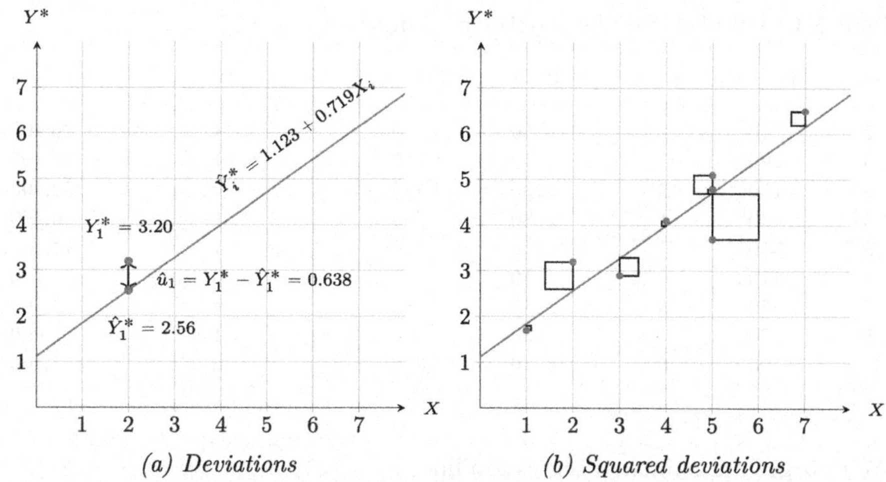

Fig. 5.3 Deviation and squared deviation between observations (Y_i^*) and the regression line $\hat{Y}_i^* = 1.123 + 0.719 \cdot X$, where $Y_i^* = Y/100$ in order to make the figure easier to read.

As shown mathematically in Sect. 5.8.1, we find the values of $\hat{\beta}_0$ and $\hat{\beta}_1$ by minimizing the sum of squared deviations between the observations and the regression line. This is illustrated in Fig. 5.3b where we see the square of each deviation, and is the reason why this method is called the ordinary least squares method. The area of a square that has four equal sides of side length $\hat{u}$ is $\hat{u}^2$, such that the squares in the figure show the magnitude of squared deviations. The sum of the deviations will be zero since the sum of the positive deviations is equal to the sum of (the absolute value of) the negative deviations. By taking the square, we will avoid this. In addition, large deviations will have an extra large effect.

Example
From our dataset with wage (Y) and education (X), where we have calculations shown in Table 3.1 (reproduced as Table 5.1) we get

$$\hat{\beta}_1 = \frac{\sum(Y_i - \bar{Y})(X_i - \bar{X})}{\sum(X_i - \bar{X})^2} = \frac{1870}{26} = 71.923$$

$$\hat{\beta}_0 = \bar{Y} - b_2\bar{X} = 400 - 71.923 \cdot 4 = 112.308$$

Hence, the slope of the regression line is 71.923 and the intersection point with the y-axis is 112.308. We also see this from the regression line in Figs. 5.1 and 5.4, where the line intersects the y-axis at 112.308 and the slope coefficient is 71.923. As shown in Fig. 5.4, we see the intersection point and the slope coefficient of the line that corresponds to $\hat{\beta}_0$ and $\hat{\beta}_1$ that we calculated.

Table 5.1 Calculation of the correlation coefficient

X	Y	$X-\bar{X}$	$Y-\bar{Y}$	$(X-\bar{X})(Y-\bar{Y})$	$(X-\bar{X})^2$	$(Y-\bar{Y})^2$
2	320	−2	−80	160	4	6,400
3	290	−1	−110	110	1	12,100
1	170	−3	−230	690	9	52,900
5	480	1	80	80	1	6,400
5	510	1	110	110	1	12,100
4	410	0	10	0	0	100
7	650	3	250	750	9	62,500
5	370	1	−30	−30	1	900
32	3,200	0	0	1,870	26	153,400

We thus get the estimated regression line

$$\hat{Y}_i = 112.3 + 71.9X_i \tag{5.7}$$

5.1.3 *Interpreting the Estimated Coefficients*

In general, we have the following interpretation of the estimated coefficients in a simple regression model:

Interpretation of constant term and slope coefficient

The slope coefficient ($\hat{\beta}_1$) shows how Y changes if X increases by 1: If X increases by 1, we expect Y to change by $\hat{\beta}_1$.

The constant term ($\hat{\beta}_0$) shows what the value of Y is expected to be if $X = 0$: If $X = 0$, Y is expected to be equal to the value of $\hat{\beta}_0$.

Since we are interested in economic interpretations of estimated coefficients, and not just what they say about the location of a straight line, we must interpret the estimates in terms of the variables in the dataset we use. In our example, we have that the dependent variable (Y) is the weekly wage measured in euros and that the independent variable (X) is the number of years of higher education. However, we should be cautious when interpreting a slope coefficient as an *effect*, since a simple linear regression is, by construction, just a recalculation of a correlation between two variables where we assume that one is the dependent and one is the independent variable.

Example
The slope coefficient expresses the change in Y if X increases by one. In our example, we thus have that:

- If the number of years of higher education increases by one year, the wage is expected to increase by EUR 71.923
- The estimated constant term $\hat{\beta}_0 = 112.308$ tells us that the expected weekly wage for a person without higher education is EUR 112.31.

In Fig. 5.4, we see how $\hat{\beta}_0$ and $\hat{\beta}_1$ correspond to the intersection point and slope of the regression line for wage and education.

A person without any higher education ($X_i = 0$) is thus expected to have a weekly wage of EUR 122.31, and if the education increases by one year, we expect the wage to increase by EUR 71.92. We can also interpret $\hat{\beta}_1$ as comparing two people, where one has an additional year of education than the other, are expected to have a wage difference of EUR 71.92 (where the person with the highest education has the highest wage since $\hat{\beta}_1 > 0$).

Often, it does not make sense to interpret the constant term as the value of Y if $X = 0$. If $X = 0$ is far away from our observed values of X or if the associated value of Y is far outside the observation range. The estimate $\hat{\beta}_0$ is then only used to place the regression line optimally in order to minimize the squared deviations.

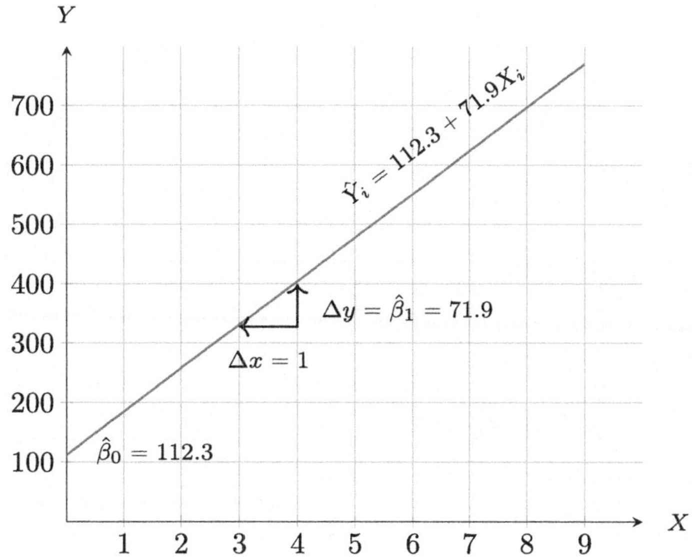

Fig. 5.4 Regression line

5.2 Simple Regression in Gretl

We can easily perform simple regression in Gretl by clicking the button marked $\hat{\beta}$ (OLS model) on the lower toolbar after we have loaded the dataset.[1] Optionally, we can choose `Model > Ordinary Least Squares`. We then get a window where we can choose which variables should be included in the model (Fig. 5.5), "const" is included by default, which means that β_0 is calculated. If we remove "const", the regression will be carried out by forcing the regression line to go through the origin since β_0 is then set equal to zero.

After selecting wage as dependent variable (`Dependent variable`) and edu as dependent variable (`Regressors`) and clicking OK, we get an output with various numbers as shown in Fig. 5.6.

We can also do this via the command line in Gretl which we bring up by pressing Crtl+S or clicking the button "open Gretl console" on the toolbar at the bottom of the main window. We can then write the line given in Code 5.1.

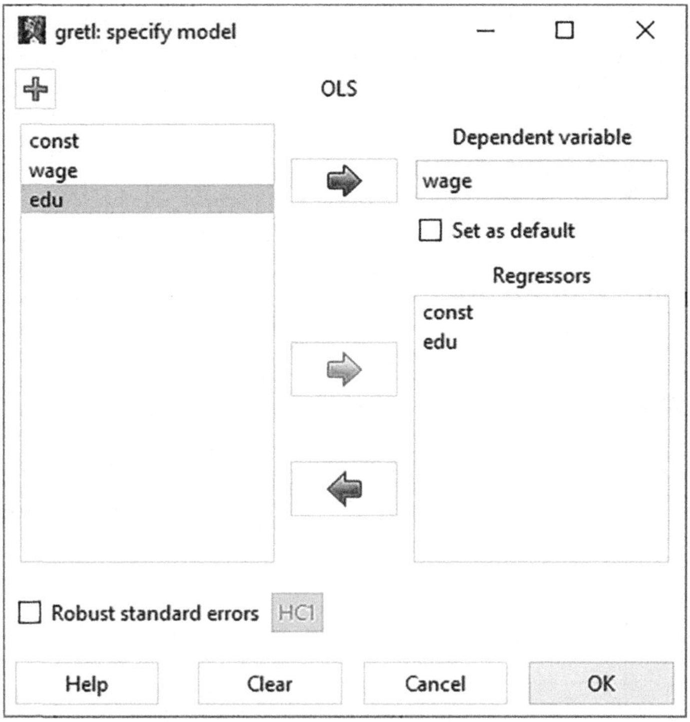

Fig. 5.5 Model specification

[1] All datasets used in the book can be downloaded from the book's website.

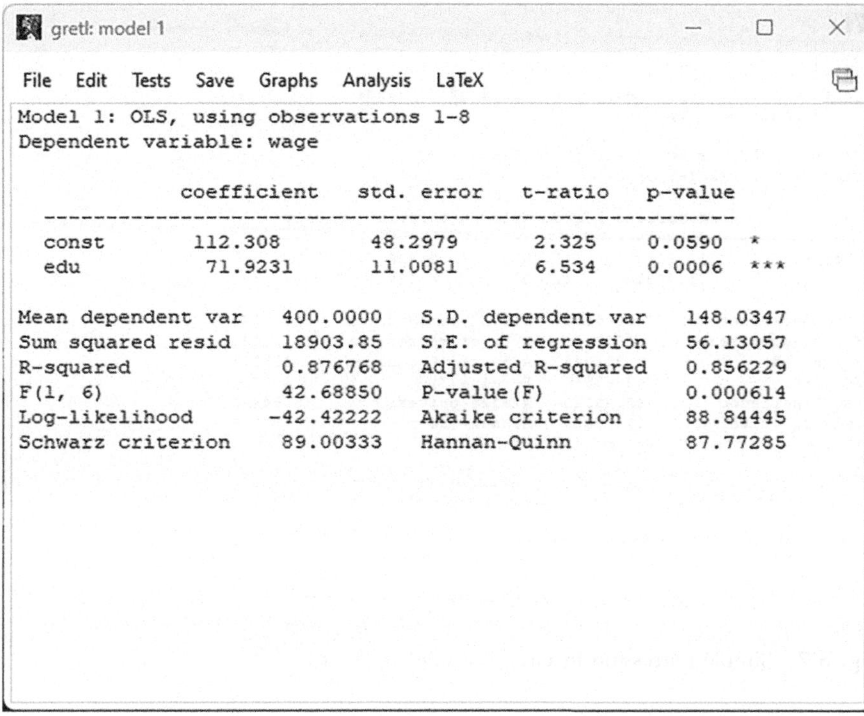

```
gretl: model 1
File Edit Tests Save Graphs Analysis LaTeX

Model 1: OLS, using observations 1-8
Dependent variable: wage

             coefficient   std. error   t-ratio   p-value
  ---------------------------------------------------------
  const        112.308       48.2979      2.325    0.0590   *
  edu           71.9231      11.0081      6.534    0.0006   ***

Mean dependent var   400.0000   S.D. dependent var   148.0347
Sum squared resid    18903.85   S.E. of regression   56.13057
R-squared            0.876768   Adjusted R-squared   0.856229
F(1, 6)              42.68850   P-value(F)           0.000614
Log-likelihood      -42.42222   Akaike criterion     88.84445
Schwarz criterion    89.00333   Hannan-Quinn         87.77285
```

Fig. 5.6 Simple regression Gretl output

Gretl code 5.1: Simple regressjon

```
ols wage const edu
```

We will then get the same output as shown in Fig. 5.6, see Fig. 5.7.

The first column (without heading) shows "const" and "edu", and refers respectively to the constant term $\hat{\beta}_0$ and the slope coefficient $\hat{\beta}_1$, while the second column shows the estimated values for β_0 and β_1; $\hat{\beta}_0$ and $\hat{\beta}_1$. We see that this corresponds to what we calculated above ($\hat{\beta}_0 = 112{,}308$ and $\hat{\beta}_1 = 71{,}923$).

By selecting `Edit` > `Copy` (or highlighting the text and pressing Ctrl+c), we can copy this table, e.g., to Microsoft Word or as plain text, and then paste it into a document where we want to write about results. We will look at the next columns in more detail after the next section.

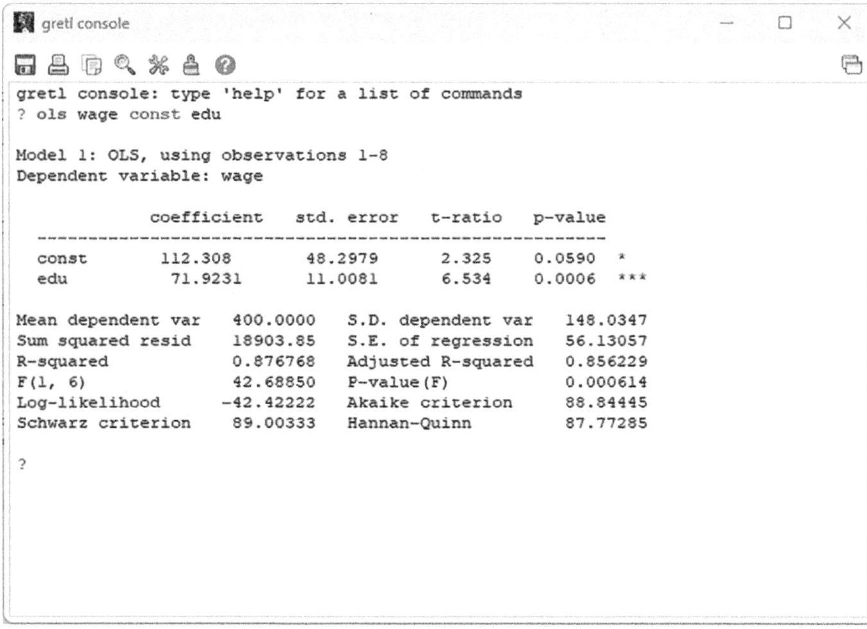

Fig. 5.7 Simple regression in Gretl console

5.3 Simple Regression in R

Regression can be performed in R by the function `lm(y ~ x)`, where the dependent variable y and the explanatory variable x are specified. If wage is the dependent variable and education is the explanatory variable, we can write `lm(wage ~ edu)`, which gives

```
Call:
lm(formula = wage ~ edu)

Coefficients:
(Intercept)    edu
112.31         71.92
```

Hence, we get the estimated coefficients $\hat{\beta}_0 = 112.31$ and $\hat{\beta}_1 = 71.92$. We can also get a full summary of the regression using the `summary()` function. By calling the regression, e.g., `singlereg` and then using the summary function we do the following:

R code 5.1: Simple regression i R

```
wage <- c(320,290,170,480,510,410,650,370)
edu <- c(2,3,1,5,5,4,7,5)
singlereg <- lm(wage ~ edu)
summary(singlereg)
```

This provides the following output in R:

```
Call:
lm(formula = wage ~ edu)

Residuals:
     Min       1Q   Median      3Q     Max
-101.923  -20.192    9.038  35.192  63.846

Coefficients:
            Estimate Std. Error t value Pr(>|t|)
(Intercept)   112.31      48.30   2.325 0.059018 .
edu            71.92      11.01   6.534 0.000614 ***
---
Signif. codes:
0 '***' 0.001 '**' 0.01 '*' 0.05 '.' 0.1 ' ' 1

Residual standard error: 56.13 on 6 degrees of freedom
Multiple R-squared:  0.8768,Adjusted R-squared:  0.8562
F-statistic: 42.69 on 1 and 6 DF,  p-value: 0.000614
```

Note that the number of stars/asterisks in the output from R represents a different level of significance than the stars/asterisks in the output from Gretl.

5.4 Prediction

We can use the estimated model to predict values of the dependent variable for a given value of the independent variable. E.g., we have that if $X = 2$, then the predicted value of Y is $\hat{Y} = \hat{\beta}_0 + \hat{\beta}_1 \cdot 2$, so we insert values of X in (5.4) to predict Y.

In our example, predicted Y if $X = 2$ will be

$$\hat{Y} = 112.308 + 71.923 \cdot 2 = 256.154$$

A person with two years of education is therefore expected to have a wage of EUR 256. For a person with 6 years of education, we expect

$$\hat{Y} = 112.308 + 71.923 \cdot 6 = 543.7$$

(a wage of EUR 543.7) a value for the variable education for which we have no observations in our dataset.

The first person in our dataset ($i = 1$) had $X = 2$ years of education, but an actual wage of EUR 320 ($Y_1 = 320$). That is, the prediction error/error term for person $i = 1$ is $\hat{u} = Y - \hat{Y} = 320 - 256.15 = 63.85$. As shown in Fig. 5.8, we see that the prediction error is the deviation between the observation (Y) and predicted value $\hat{Y}$. The predicted value is the value of Y that corresponds to the regression line.

We can do this for all eight people and thus get predicted values as shown in Table 5.2.

One thing we should notice is that persons 4, 5, and 8 have a predicted wage of EUR 471.92 since all of these have an education of $X_i = 5$ years. If $X = 5$ then we get $\hat{Y} = 471.92$ as shown by the regression line. We can also generate the predicted values as the variable `yhat=` $\hat{Y}_i$ and the error terms as `uhat=` $\hat{u}_i$ by the commands `$yhat` and `$uhat` in Gretl given in Gretl code 5.2 or by select `Save > Fitted Values` and `Save > Residuals` after estimating the model in Gretl.

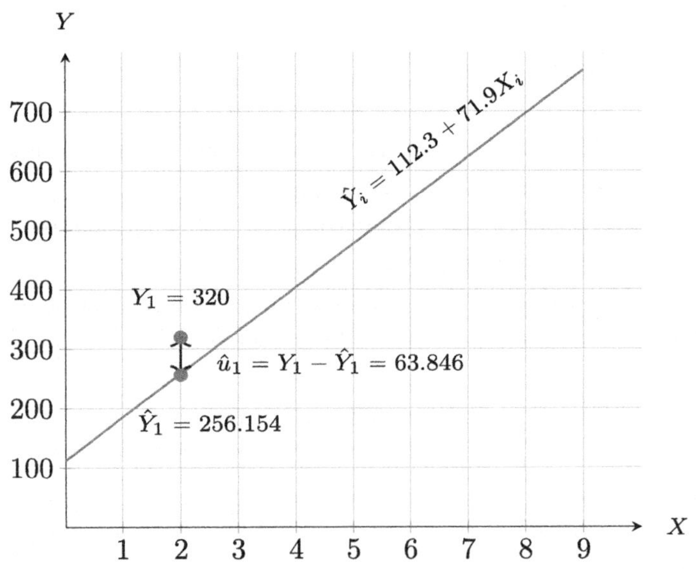

Fig. 5.8 Regression line and residual

Table 5.2 Observations and predicted dependent variable

i	X	Y	$\hat{Y}$
1	2	320	256.15
2	3	290	328.08
3	1	170	184.23
4	5	480	471.92
5	5	510	471.92
6	4	410	400.00
7	7	650	615.77
8	5	370	471.92
SUM	32	3,200	3,200.00

Gretl code 5.2: Generating predicted values and error terms

```
yhat=$yhat
uhat=$uhat
```

In R, we can calculate predicted values and residual duals, $\hat{Y}_i$ and $\hat{u}_i$, as follows:

R code 5.2: Generating predicted values and error terms

```
yhat <- fitted(singlereg)
uhat <- resid(singlereg)
```

5.5 Explanatory Power of the Model

Predicted value $\hat{Y}$ (as we saw in Table 5.2) is the value given by the regression line, and $\hat{u}$ is the deviation between the regression line and the observed value (also called residual or error term. We see this in Fig. 5.3 where the points along the straight line for our observed values of X provide $\hat{Y}$, and the vertical distance between the line and the observation (the points) gives the deviations $\hat{u}$.

The sum of the deviations becomes zero since the line is placed and the squared deviations ($\widehat{u^2}$) are minimized. We expand Table 5.2 to include deviation and the sum of the squared deviations $\sum(Y - \hat{Y})^2$. In addition, we include squared deviation between Y and $\bar{Y}$ and between $\hat{Y}$ and $\bar{Y}$, which is useful to tell us how good the model is.

In Table 5.3, we have calculated three different deviations, the square of these deviations, and the sum of all columns. These three sums can be interpreted as follows:

Table 5.3 Predicted dependent variable, error term, and squared error term

i	X	Y	$\hat{Y}$	$Y-\hat{Y}$	$(Y-\hat{Y})^2$	$Y-\bar{Y}$	$(Y-\bar{Y})^2$	$\hat{Y}-\bar{Y}$	$(\hat{Y}-\bar{Y})^2$
1	2	320	256.15	63.85	4,076.33	−80	6,400	−143.85	20,691.72
2	3	290	328.08	−38.08	1,449.85	−110	12,100	−71.92	5,172.93
3	1	170	184.23	−14.23	202.51	−230	52,900	−215.77	46,556.36
4	5	480	471.92	8.08	65.24	80	6,400	71.92	5,172.93
5	5	510	471.92	38.08	1,449.85	110	12,100	71.92	5,172.93
6	4	410	400	10.00	100	10	100	0.00	0.00
7	7	650	615.77	34.23	1,171.75	250	62,500	215.77	46,556.36
8	5	370	471.92	−101.92	10,388.31	−30	900	71.92	5,172.93
	32	3,200	3,200	0	18,903.85	0	153,400	0	134,496.15

"Total sum of squares":

$$\text{TSS} = \sum_{i=1}^{n} (Y_i - \bar{Y})^2 = 153400$$

This is the total variation in Y that we calculated for the standard deviation of Y. It does not say anything about the model, but tells how much Y varies in relation to its mean.

- This can be found in Gretl by selecting `Analysis > ANOVA` in the window with the output from the regression. If we use the command line, we can write `ols wage const edu -anova` as shown in Code 5.3 to get the same table. We can find it from row `Total` in column `Sum of squares` in Fig. 5.9, as the value 153,400.
- In R we can calculate this by writing `tss <- sum((wage-mean(wage)^2)` if wage is the dependent variable. This provides the value 153,400.

"Residual sum of squares":

$$\text{RSS} = \sum_{i=1}^{n} (Y_i - \hat{Y}_i)^2 = \sum_{i=1}^{n} (Y_i - b_1 - b_2 X_i)^2 = \sum_{i=1}^{n} u_i^2 = 18903.85$$

This is the sum of the squared deviations between the regression line and observed values. Through the use of the ordinary least squares method and the formulas for the coefficients in the regression model, this becomes as small as possible.

- In Gretl we find we find the value in the ANOVA table in the row `Residual`, and in the output from the regression as `Sum squared resid`.
- In R we can bring up the ANOVA table by typing `anova <- aov(singlereg)` and `summary(anova)`. The value for RSS is in the "Residuals" row in the "Sum Sq" column. The table is shown below.

```
           Df Sum Sq Mean Sq F value   Pr(>F)
edu         1 134496  134496   42.69 0.000614 ***
Residuals   6  18904    3151
---
Signif. codes:
0 '***' 0.001 '**' 0.01 '*' 0.05 '.' 0.1 ' ' 1
```

"Explained sum of squares":

$$\text{ESS} = \sum_{i=1}^{n}(\hat{Y}_i - \bar{Y})^2 = \sum_{i=1}^{n}(b_1 + b_2 X_i - \bar{Y})^2 = 134496.15$$

This sum shows the deviation between predicted values for Y and the mean of Y. This is thus the variation in Y that is explained by the model.

- In Gretl, we can find this value in the ANOVA table for the row `Regression`.
- In R we find the value for ESS in the row "edu" in the column "Sum Sq" in the ANOVA table.

Gretl code 5.3: Analysis of variance (ANOVA) in Gretl

```
ols wage const edu --anova
```

R code 5.3: Analysis of variance (ANOVA) in R

```
anova <- aov(enkelreg)
summary(anova)
sum((wage-mean(wage))^2) #For calculating TSS
```

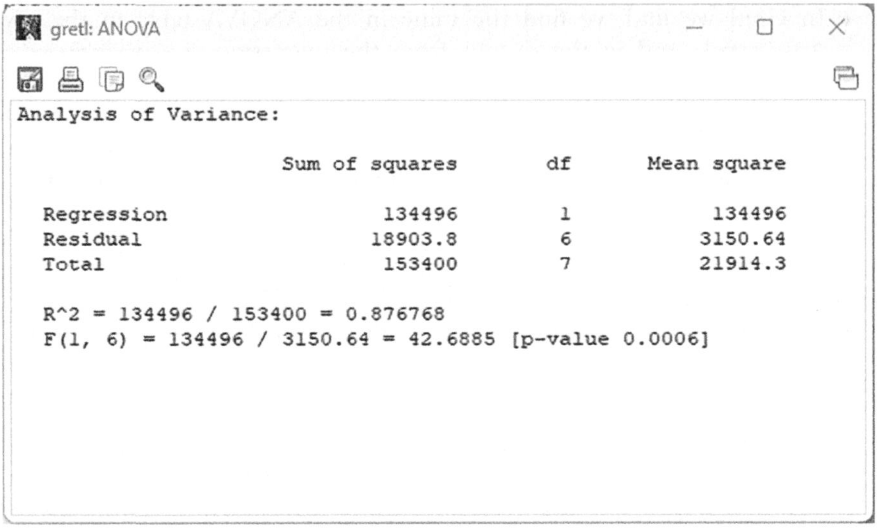

Fig. 5.9 Analysis of variance (ANOVA) in Gretl

The share of the variation in the dependent variable that is explained by the model relative to the total variation in the dependent variable is called **R-squared**, R^2. This can thus be calculated as

$$R^2 = \frac{\text{ESS}}{\text{TSS}} = \frac{\sum_{i=1}^{n}\left(\hat{Y}_i - \bar{Y}\right)^2}{\sum_{i=1}^{n}\left(Y_i - \bar{Y}\right)^2} \tag{5.8}$$

R-squared

R-squared (R^2) shows the share of the variation in Y that is explained by the model and is thus a measure of how good the model is.

Since

$$\text{TSS} = \text{ESS} + \text{RSS}$$

(the sum of explained and unexplained variation constitutes total variation), we can also calculate R-squared as

$$R^2 = \frac{\text{ESS}}{\text{TSS}} = \frac{\text{TSS} - \text{RSS}}{\text{TSS}} = 1 - \frac{\text{RSS}}{\text{TSS}} \tag{5.9}$$

We thus need TSS as well as either RSS or ESS (but not necessarily both) to calculate R^2.

In our example, we get

$$R^2 = \frac{\text{ESS}}{\text{TSS}} = \frac{134496.15}{153400} = 0.877$$

or

$$R^2 = 1 - \frac{\text{RSS}}{\text{TSS}} = 1 - \frac{18903.85}{153400} = 1 - 0.123 = 0.877$$

Hence, 87.7% of the variation in Y (wage) is explained by the model $\hat{Y} = 112.308 + 71.923 \cdot X$ (i.e., a linear model that depends on education) while 12.3% of the variation is explained by other factors that are not included in the model. These can be non-linear effects, experience, work tasks, productivity, etc. The value is given in the Gretl output as `R-squared` and in the R output as `Multiple R-squared`.

Adjusted R-squared is another measure of the model's explanatory power that we use if we compare several models that try to say something about a variable (i.e., models with the same dependent variable), but that have different numbers of explanatory variables. This can be calculated as

$$\bar{R}^2 = 1 - \left[\left(1 - R^2\right) \cdot \left(\frac{n-1}{n-k-1}\right)\right] \tag{5.10}$$

where k is the number of increments. In our example, we thus get

$$\bar{R}^2 = 1 - \left[(1 - 0.877) \cdot \left(\frac{8-1}{8-1-1}\right)\right] = 1 - \left[0.123 \cdot \frac{7}{6}\right] = 0.8565$$

since we have $k = 1$ explanatory variable (education). We find this value in the Gretl output and in the R output (slight deviation due to rounding of numbers) as `Adjusted R-squared`.

We can also calculate **the standard deviation and the variance of the regression model** which says something about the spread of the error terms. The standard deviation is estimated as

$$\hat{\sigma} = \sqrt{\frac{\sum \hat{u}_i^2}{n-k-1}} \tag{5.11}$$

We can also write this as

$$\hat{\sigma} = \sqrt{\frac{\sum \hat{u}_i^2}{n-k-1}} = \sqrt{\frac{\sum_{i=1}^{n}(Y_i - \hat{Y}_i)^2}{n-k-1}} = \sqrt{\frac{\text{RSS}}{n-k-1}} \tag{5.12}$$

The larger the standard deviation, the poorer the explanatory power of the model. In our example, we get

$$\hat{\sigma} = \sqrt{\frac{\text{RSS}}{n-k-1}} = \sqrt{\frac{18904}{8-2}} = 56.13$$

We find this value in the Gretl output as `S.E. of regression` and in the R output as `Residual standard error`. We also have that the variance is

$$\hat{\sigma}^2 = \frac{\text{RSS}}{n-k-1} = \frac{18904}{8-2} = 3150.64$$

The variance or standard deviation of the model does provide much information on its own, therefore we usually use R^2 since it lies between 0 and 1 and expresses the explanatory power as a share. However, we need the model's standard deviation to calculate the precision of the estimated coefficients, which we will do in the next subsection.

5.6 Hypothesis Testing and Confidence Intervals

5.6.1 Hypothesis Testing of the Slope Coefficient

We can test whether there is a statistically significant relationship between the variables by performing a hypothesis test for the slope coefficient:

$$H_0 : \beta_1 = 0 \text{ (There is no relationship between X and Y)}$$
$$H_A : \beta_1 \neq 0 \text{ (There is a relationship between X and Y)}$$

Hence, we have a two-sided hypothesis test, where we see that if $\beta_1 = 0$ there will be no linear relationship between X and Y.

Usually, we do not know the population's standard deviation σ, so we use the t-distribution with $n - k - 1$ degrees of freedom (k refers to the number of slope coefficients or explanatory variables in the model which is $k = 1$ in simple regression as here).

The test statistic when conducting a hypothesis test on a slope coefficient is

$$T = \frac{\hat{\beta}_1}{\text{SE}(\hat{\beta}_1)} \tag{5.13}$$

Also note that if we have a different value than zero in the null hypothesis, we need to use $T = \frac{\hat{\beta}_1 - \beta_1^{H0}}{\text{SE}(\hat{\beta}_1)}$, where β_1^{H0} is the value in the null hypothesis. E.g., if we have $H_0 : \beta_1 = 1$ and $H_A : \beta_1 \neq 1$, we get $T = \frac{\hat{\beta}_1 - 1}{\text{SE}(\hat{\beta}_1)}$.

While $\hat{\beta}_1$ is the estimated slope coefficient of the model, $\text{SE}(\hat{\beta}_1)$ is the standard error of this estimate which tells us something about how precise the estimate is. If the standard error is large, there is a large statistical uncertainty associated with the value of the estimated value $\hat{\beta}_1$. In simple regression, the standard error is calculated as

$$SE(\hat{\beta}_1) = \sqrt{\frac{\hat{\sigma}^2}{\sum_{i=1}^{n}(X_i - \bar{X})^2}} \tag{5.14}$$

where σ^2 is the model's estimated variance corresponding to the square of the standard deviation shown in (5.12).

We can find the critical value in the t-distribution (if we do not know the population's standard deviation) with $n - k - 1$ degrees of freedom for the chosen significance level α. Remember to use $\alpha/2$ as tail probability and two critical values $\pm t_{\alpha/2}(n - k - 1)$ if you have a two-sided test.

Hypothesis testing for slope coefficients

Approach:

1) Set up the hypotheses (where we often have $H_A : \beta_1 = 0$ which we would like to test)
2) Calculate test statistic $T = \hat{\beta}_1/SE(\hat{\beta}_1)$
3) Find critical value(s) ($\pm t_{\alpha/2}(n - k - 1)$ for two sided test and $t_\alpha(n - k - 1)$ or $-t_\alpha(n - k - 1)$ for one sided test)
4) Reject the null hypothesis if the test statistic is in the rejection area ($|T| > k$) *or*

   Keep the null hypothesis if the test statistic is not in the rejection area ($|T| \leqslant k$)

Example

In our example, we previously found that $\hat{\sigma}^2 = 3150.64$ and $\sum_{i=1}^{n}(X_i - \bar{X})^2 = 26$. We can thus calculate the standard error of $\hat{\beta}_1$ as

$$SE(\hat{\beta}_1) = \sqrt{\frac{\hat{\sigma}^2}{\sum_{i=1}^{n}(X_i - \bar{X})^2}} = \sqrt{\frac{3150.64}{26}} = \sqrt{121.178} = 11.01$$

This is shown in the column `std. error` in the Gretl output and in the column `Std. Error` in R. Our estimate of the population effect of education on wage was $\hat{\beta}_1 = 71.92$ (one year of additional education leads to an expected wage increase of EUR 71.92), and a typical deviation is thus the standard error of EUR 11.01.

If we for the model $Y_i = \beta_0 + \beta_1 X_i + u_i$ we have the hypotheses

$$H_0 : \beta_1 = 0 \text{ (education does not affect wages)}$$
$$H_A : \beta_1 \neq 0 \text{ (education affects wages)}$$

we can calculate the test statistic and find critical values in order to perform the hypothesis test.

The test statistic becomes

$$T = \frac{\hat{\beta}_1}{\mathrm{SE}(\hat{\beta}_1)} = \frac{71.92}{11.01} = 6.53$$

This is shown in the `t-ratio` column in the Gretl output and the `t value` column in R. Note that this is similar to the test statistic we found when testing for a significant correlation between the two variables in Sect. 4.4, since simple regression is by construction an alternative way of estimating a correlation.

Critical values are found in the t-distribution with $n - k - 1$ degrees of freedom. If we choose a 5% significance level, we thus have

$$\pm t_{\alpha/2}(n - k - 1) = \pm t_{0.025}(6) = \pm 2.447$$

The test observer is located in the rejection area. Hence, we reject the null hypothesis, which implies that we find that education has an effect on wages. We can also see this directly from the p-value (in the column `p-value` in the output from Gretl and in the column `Pr(>|t|)` in R) which is 0.0006 and thus less than the significance level of 0.05 (5%). For Gretl, the three asterisks (star-shaped characters ***) in the rightmost column show that the effect is significant at a 1% significance level, while two (**) and an (*) asterisk shows respectively significance at a 5% and 10% significance level. In R are used respectively two asterisks, an asterisk, and a dot for the same (see description in row `Signif. codes` in the R output).

5.6.2 *Confidence Interval for a Slope Coefficient*

Since we have the standard error and the estimate of a coefficient, we can also calculate **the confidence interval of the coefficient**.

> **Confidence interval for a slope coefficient**
>
> $(1 - \alpha) \cdot 100\%$ of confidence intervals will contain the population parameter at a α significance level.

We can calculate this as

$$\hat{\beta}_i \pm t_{\alpha/2}(n - k - 1) \cdot \mathrm{SE}(\beta_i) \tag{5.15}$$

A 95% confidence interval for β_1 in our example thus becomes

$$\hat{\beta}_1 \pm t_{0.025}(6) \cdot \mathrm{SE}(\beta_1) = 71.92 \pm 2.447 \cdot 11,01 = [44.99,\ \ 98.86]$$

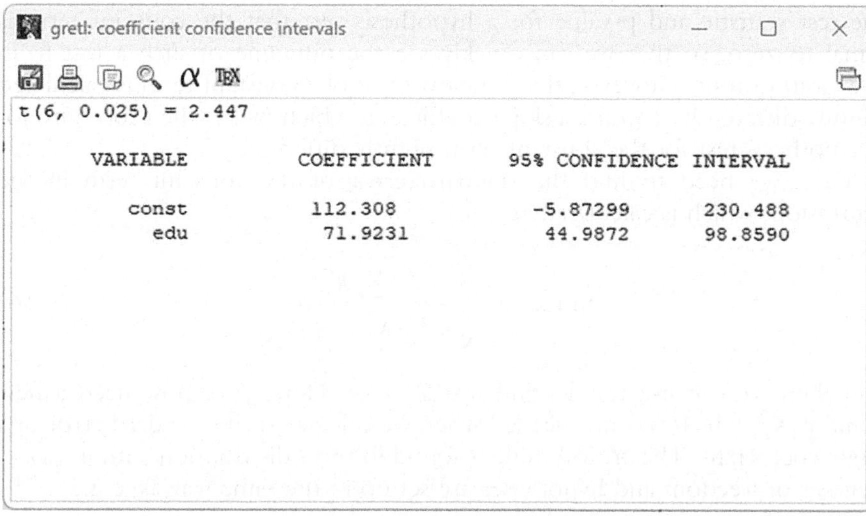

Fig. 5.10 Confidence interval for coefficients in Gretl

We can also use this to carry out the hypothesis test above. The null hypothesis value $\beta_1 = 0$ does not lie in the interval, so that we can thus reject the null hypothesis $\beta_1 = 0$ at a 5% significance level.

In Gretl, we can find the coefficients' confidence interval by choosing `Analysis > Confidence interval for coefficients`. In our example, we then get (Fig. 5.10).

By clicking on the button labeled α (Confidence level...) we can choose a confidence interval other than the default 95%. The output also shows the confidence interval for the constant term β_0.

In R we can write `confint(singlereg)` to find the 95% confidence interval for the coefficients. This gives

```
                2.5 %    97.5 %
(Intercept) -5.872986 230.48837
edu         44.987200  98.85895
```

We can also choose a different confidence interval by adding the argument `level`. For instance, `confint(singlereg,level = 0.90)` will provide a 90% confidence interval (95% is the default setting).

5.6.3 *Hypothesis Test for the Constant Term*

We can also perform a hypothesis test about the constant term in a regression model. This may not be something you need to do as often as for slope coefficient, but it can be useful in some cases. The output from Gretl gives the test statistic and p-value for a hypothesis test that the constant term is equal to zero, so that we can quickly see the outcome of such a test from the Gretl output. However, the standard error of a constant term is calculated slightly differently than for a slope coefficient, which makes the procedure for a hypothesis test for the constant term slightly different.

First, we need to find the standard error of the constant term in the regression, which is calculated as

$$\mathrm{SE}(\hat{\beta}_0) = \sqrt{\frac{\hat{\sigma}^2 \sum X_i^2}{n \sum (X_i - \bar{X})^2}} \tag{5.16}$$

and then we can use this to find test statistic. Hence, we now need a new sum, $\sum X_i^2$, which was not needed when we calculated the standard error of a slope coefficient. The critical value is found in the t-distribution with $n - k - 1$ degrees of freedom and hypotheses are set up in the same way as, e.g.,

$$H_0 : \beta_0 = 0 \text{ (the constant term is zero)}$$
$$H_A : \beta_0 \neq 0 \text{ (the constant term is not equal to zero)}$$

In our example, we then need to calculate

$$\sum_{i=1}^{n} X_i^2 = \sum_{i=1}^{8} X_i^2 = 2^2 + 3^2 + 1^2 + 5^2 + 5^2 + 4^2 + 7^2 + 5^2 = 154$$

which yields

$$\mathrm{SE}(\hat{\beta}_0) = \sqrt{\frac{3150.64 \cdot 154}{8 \cdot 26}} = \sqrt{\frac{485198.56}{208}} = \sqrt{2332.685} = 48.299$$

We can find this value from the Gretl output in the column `std. error` for the constant term `const`, or from the R output in the column `Std. Error` for the row `(Intercept)` (somewhat different due to rounding).

We can then calculate the test statistic

$$T = \frac{\hat{\beta}_0}{\mathrm{SE}(\hat{\beta}_0)} = \frac{112.308}{48.299} = 2.325$$

Critical values at a 5% significance level are the same as for the hypothesis test we did for the slope coefficient

$$\pm t_{\alpha/2}(n - k - 1) = \pm t_{0.025}(6) = \pm 2.447$$

The conclusion is thus that we keep the null hypothesis at a 5% significance level since the test statistic is not in the rejection area. However, we see from the Gretl and R outputs that the p-value is 0.059. That is, we can reject the null hypothesis at a 5.9% significance level or higher. This is also shown by * which indicates that we can reject the null hypothesis if we have a 10% significance level.

5.6.4 *Prediction Interval*

Just as we can create a confidence interval for an estimated coefficient that says something about the uncertainty when estimating the population coefficient (see Sect. 5.6.2), we can create an interval for a predicted value of the dependent variable. We often want to use a regression model to predict what the dependent variable will be, so it can be useful to also show the uncertainty related to the predicted value through. The prediction interval is defined as

$$\hat{Y} \pm t_{\alpha/2}(n-k-1) \cdot \hat{\sigma} \cdot \sqrt{\frac{n+1}{n} + \left(\frac{X-\bar{X}}{\hat{\sigma}/\mathrm{SE}(\hat{\beta}_1)}\right)^2} \tag{5.17}$$

where X_i is the value of X that we want to use to predict Y. If $X = 2$, then we get $\hat{Y} = 256.15$ as shown, e.g., in Table 5.3. A 95% prediction interval then becomes

$$256.15 \pm t_{0.025}(6) \cdot 56.13 \cdot \sqrt{\frac{9}{8} + \left(\frac{2-4}{56.13/11.01}\right)^2}$$

$$256.15 \pm 2.447 \cdot 56.13 \cdot \sqrt{1.125 + \left(\frac{-2}{5.098}\right)^2}$$

$$256.15 \pm 2.447 \cdot 56.13 \cdot \sqrt{1.279}$$

$$256.15 \pm 2.447 \cdot 56.13 \cdot 1.131$$

$$256.15 \pm 155.34$$

which provides the interval

$$[100.81 \ , \ 409.49]$$

This means that if $X = 2$, Y we get a confidence interval between 100.81 and 409.49. 95% of confidence intervals will contain the true value of Y at a 5% significance level.

Prediction Interval in Gretl and R

In order to calculate a prediction interval in Gretl, you can go to `Analysis` > `Forecasts...` after the regression is carried out. Leave everything as it is by default, but you may change the confidence interval if desired. This gives (Fig. 5.11).

We get the output (Fig. 5.12).

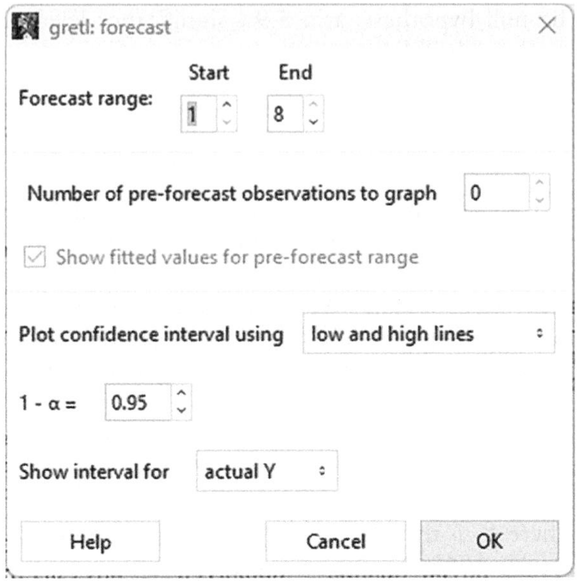

Fig. 5.11 Prediction interval in Gretl

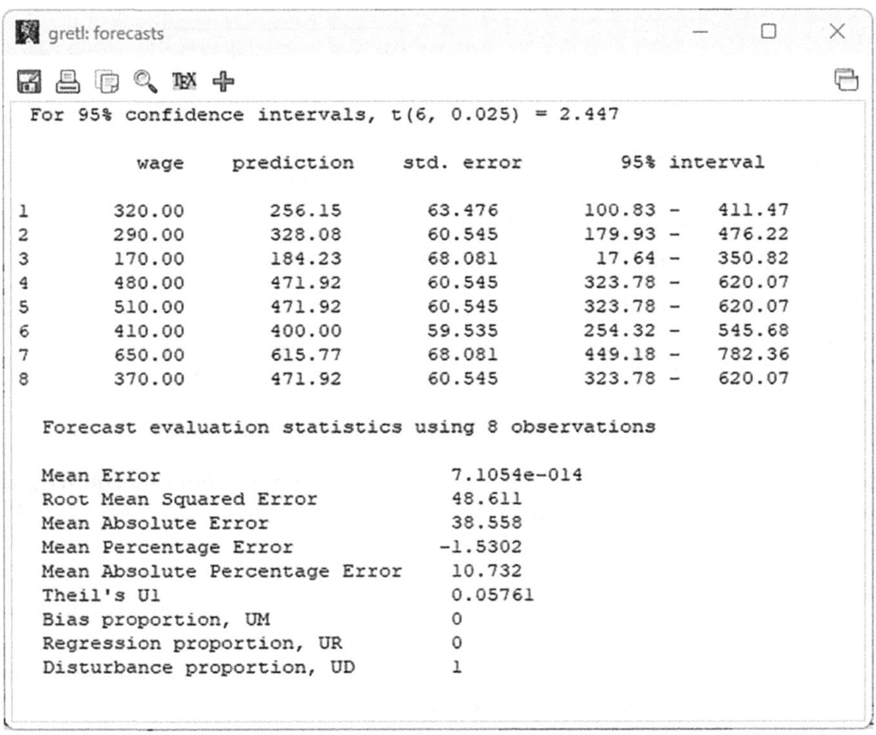

gretl: forecasts

For 95% confidence intervals, t(6, 0.025) = 2.447

	wage	prediction	std. error	95% interval
1	320.00	256.15	63.476	100.83 - 411.47
2	290.00	328.08	60.545	179.93 - 476.22
3	170.00	184.23	68.081	17.64 - 350.82
4	480.00	471.92	60.545	323.78 - 620.07
5	510.00	471.92	60.545	323.78 - 620.07
6	410.00	400.00	59.535	254.32 - 545.68
7	650.00	615.77	68.081	449.18 - 782.36
8	370.00	471.92	60.545	323.78 - 620.07

Forecast evaluation statistics using 8 observations

Mean Error	7.1054e-014
Root Mean Squared Error	48.611
Mean Absolute Error	38.558
Mean Percentage Error	-1.5302
Mean Absolute Percentage Error	10.732
Theil's U1	0.05761
Bias proportion, UM	0
Regression proportion, UR	0
Disturbance proportion, UD	1

Fig. 5.12 Prediction interval output in Gretl

We thus find for person $i = 1$ that a 95% prediction interval lies between 100.83 and 411.47 which is in line with what we calculated for $X = 2$ above (some difference due to rounding).

We also get a figure showing the intervals (Fig. 5.13).

From the figure, we see that the prediction interval is larger the further away from the mean of X ($\bar{X} = 4$) we get. This is because the predictions for Y are more precise closer to the mean of X.

In R, we can find the 95% prediction interval by typing `predict(singlereg, interval="predict")`. We then get the predicted value, as well as the lower and upper limits of the interval (`lwr` and `upr`). By adding the option `level` we can choose a different interval, e.g., `predict(singlereg, interval="predict", level = 0.90)` will give a 90% prediction interval.

R code 5.4: Prediction interval in R

```
predict(singlereg, interval="predict")
```

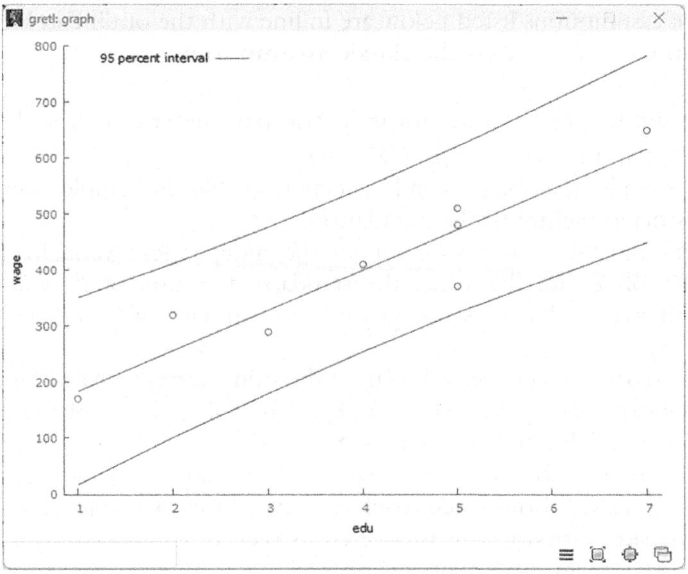

Fig. 5.13 Prediction interval, figure

For a 95% prediction interval, we get

```
       fit       lwr      upr
1 256.1538 100.83401 411.4737
2 328.0769 179.92960 476.2242
3 184.2308  17.64157 350.8200
4 471.9231 323.77576 620.0704
5 471.9231 323.77576 620.0704
6 400.0000 254.32197 545.6780
7 615.7692 449.18003 782.3584
8 471.9231 323.77576 620.0704
```

including a warning that "predictions on current data refer to _future_ responses". This warning only concerns the assumptions behind the calculated intervals since it is a confidence interval that is calculated around an *estimated* value of the dependent variable.

5.7 Assumptions for the Regression Model

For the simple regression model, there are some assumptions that must be met for our estimates to be **unbiased**. In particular, the estimators $\hat{\beta}_0$ and $\hat{\beta}_1$ for the population parameters β_0 and β_1 are unbiased, meaning that with a random sample we can expect the estimate to give us the expected population value. The assumptions listed below are in line with the outline in Wooldridge (2019), and are often called **the classic assumptions**

1. The model needs to be **linear in the parameters**, such as $Y_i = \beta_0 + \beta_1 X_i + u_i$ and not $Y_i = \beta_0 + X_i^{\beta_1} + u_i$.
2. The sample must be randomly selected, so that the sample does not give a distorted picture of the population.
3. There must be some variation in the independent variable. If all the values of X_i are the same, the standard deviation of X_i will be zero, which means that it is not possible to estimate the coefficients in the model.
4. The error term is expected to have the value zero regardless of the value of the explanatory variable: $E(u_i|X_i) = 0$. This can be violated if we have omitted variable bias, see Sect. 6.8.
5. The variance of the error term is constant: $Var(u_i|X_i) = \sigma^2$. This is called **homoskedasticity**, and in the opposite case we have **heteroskedasticity**. This is covered in Sect. 6.5.
6. The error term is normally distributed.

If the first four assumptions are met, the estimators are unbiased. If, in addition, assumption no. 5 is fulfilled, the estimators are the most efficient and "**BLUE**"; Best linear unbiased estimators. If assumption 6 also is fulfilled, we can use the t-distribution to do hypothesis tests about the coefficients.

5.7.1 Graphical Assessment of the Assumptions

The assumptions listed in Sect. 5.7 considers the population residuals u_i and not the residuals which results from the estimation over a sample, $\hat{u}_i$. Hence, most of the graphical assessment in this chapter do not tell us anything about whether the assumptions hold or not. Formal tests for some of these assumptions will be covered later in the book.

Graphical Assessment of the Assumptions in Gretl
There are various tests to formally test whether these prerequisites are met, but at this stage we will only assess this graphically.

1. `Graphs > Residual plot > Against edu`
   Used to examine whether u and x are independent. Shown in Fig. 5.14a.
2. `Graphs > Residual plot > By observation number`
   Used to check whether the error terms are independent of each other. Shown in Fig. 5.14b.
3. `Tests > Normality of Residual`
   Used to check whether the error terms are normally distributed. Shown in Fig. 5.14c.

Other useful things we can find:

- `Analysis > Display actual, fitted, residual`
  Shows all Y_i, $\hat{Y}_i$ (fitted) and $\hat{u}_i$ (residuals/error terms)
- `Graphs > Fitted, actual plot > Actual vs. fitted`
  Displays the regression line and the observed values for Y as in a scatter plot. Shown in Fig. 5.14d.
- `Save > Fitted values`
  Stores the predicted values $\hat{Y}_i$ as a new variable in the dataset.
- `Save > Residuals`
  Stores the error terms $\hat{u}_i$ as a new variable in the dataset.

As we can see in Fig. 5.14a, there does not appear to be any systematic connection between the error terms and education. If there had been a larger residual for people with high education than with low education, we would have had problems with **heterogeneity** (which is covered further in Sect. 6.5). The same applies to the error term and observation number as shown in Fig. 5.14b where there is no visible pattern in the size of the deviations. The

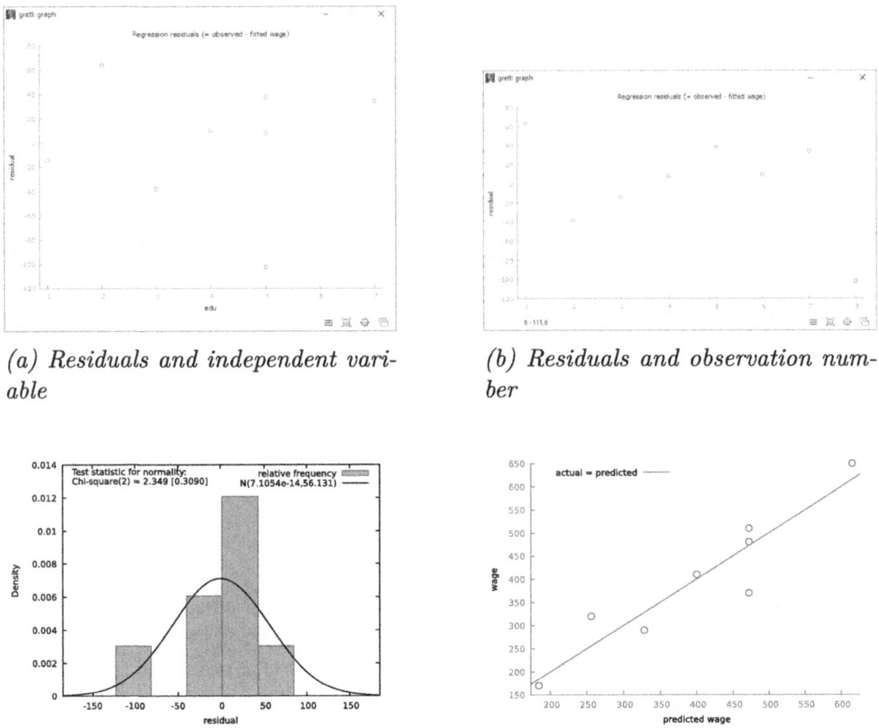

(a) Residuals and independent variable

(b) Residuals and observation number

(c) Distribution of residuals

(d) Predicted and observed values

Fig. 5.14 Graphical presentation of residuals and predicted values in Gretl

residuals also appear to be normally distributed as we can see in Fig. 5.14c, although it is difficult to judge this based on only eight observations. The figure with predicted and observed values (Fig. 5.14d) also shows this. There are also formal tests for this, where we will cover tests for heteroskedasticity in Sect. 6.5.

Graphical Assessment of the Assumptions in R

As shown in R code 5.2, we can store the residuals from a regression with the command `resid(singlereg)` and predicted variables with the function `fitted(singlereg)`, where `singlereg` is the name we have given the regression model. In order to assess graphically whether the prerequisites are met, we can create scatter plots and histograms. The codes to create the shapes are given in the R code below. `main` gives the caption of figure, while `xlab` and `ylab` gives the title on the horizontal and vertical axes, respectively, and the number on `pch` determines the appearance of the observations (`pch=19` implies a solid circle). The figures are shown in Fig. 5.15.

R code 5.5: Create figures for analyzing error terms

```
plot(edu, uhat , main="Scatterplot residuals and
independent variable", xlab="education",
ylab="residual", pch=19)
plot(uhat , main="Scatterplot residuals and
observation number", xlab="observation number",
ylab="residual", pch=19)
hist(uhat, main="Histogram residuals", xlab="residual",
ylab="frequency", pch=19)
plot(wage, yhat , main="Scatter plot wage and
predicted wage", xlab="yhat", ylab="wage", pch=19)
```

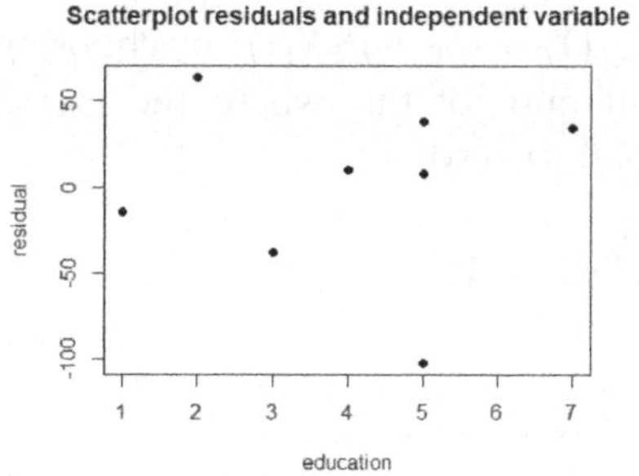

(a) Residuals and independent variable

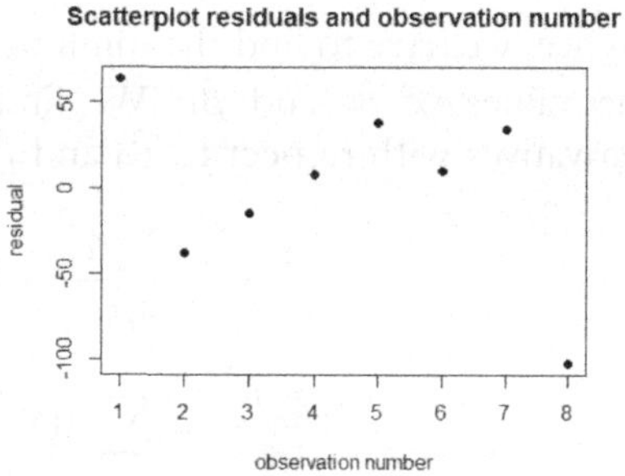

(b) Residuals and observation number

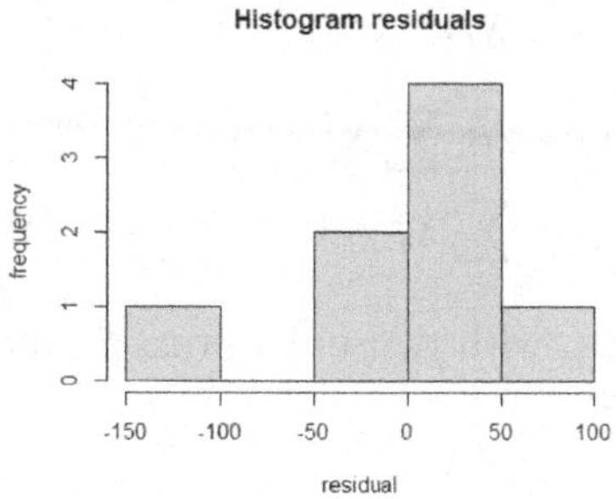

(c) Distribution of residuals

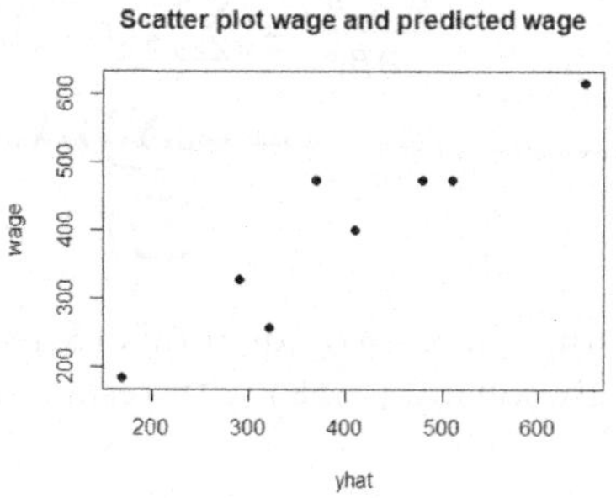

(d) Predicted and observed values

Fig. 5.15 Graphical presentation of residuals and predicted values in R

We get the same figures here as from Gretl. See the analysis according to Fig. 5.14 for assessment of the assumptions. Formal misspecification tests for heteroskedasticity are shown in Sect. 6.5.

5.8 Supplementary Material

5.8.1 Where Do the Formulas for the Coefficients Come from?

Simple regression (and multiple regression which is covered in the next chapter) is often called the ordinary *least squares* method. This is because you find the straight line that fits the observations best by minimizing the squared deviations between the line and the observations as shown in Fig. 5.3b. We must therefore minimize the sum of the squared error terms $\hat{u}_i^2$. Since these can be written as

$$\sum_{i=1}^{n} \hat{u}^2 = \sum_{i=1}^{n} \left(Y_i - \hat{Y}_i\right)^2 = \sum_{i=1}^{n} \left(Y_i - \left(\hat{\beta}_0 + \hat{\beta}_1 X_i\right)\right)^2$$

Hence, we have to find the minimum of $\sum_{i=1}^{n} \left(Y_i - \left(\hat{\beta}_0 + \hat{\beta}_1 X_i\right)\right)^2$ by choosing the values of $\hat{\beta}_0$ and $\hat{\beta}_1$. We find the minimum of this where the partial derivatives with respect to $\hat{\beta}_0$ and $\hat{\beta}_1$ are equal to zero:

$$\frac{\partial \sum \hat{u}^2}{\partial \hat{\beta}_0} = 0 \text{ and } \frac{\partial \sum \hat{u}^2}{\partial \hat{\beta}_1} = 0$$

$$\begin{aligned}
\frac{\partial \sum \hat{u}^2}{\partial \hat{\beta}_0} &= \sum 2\left(Y_i - \left(\hat{\beta}_0 + \hat{\beta}_1 X_i\right)\right) \cdot (-1) \\
&= -2\sum Y_i + 2nb_1 + 2\hat{\beta}_1 \sum X_i = 0 \\
&\quad -\sum Y_i + nb_1 + \hat{\beta}_1 \sum X_i = 0
\end{aligned} \tag{5.18}$$

$$\begin{aligned}
\frac{\partial \sum \hat{u}^2}{\partial \hat{\beta}_1} &= \sum 2\left(Y_i - \left(\hat{\beta}_0 + \hat{\beta}_1 X_i\right)\right) \cdot (-X_i) \\
&= -2\sum Y_i X_i + 2\hat{\beta}_0 \sum X_i + 2\hat{\beta}_1 \sum X_i^2 = 0 \\
&\quad -\sum Y_i X_i + \hat{\beta}_0 \sum X_i + \hat{\beta}_1 \sum X_i^2 = 0
\end{aligned} \tag{5.19}$$

We thus have two equations (5.18) and (5.19), with two unknowns, $\hat{\beta}_0$ and $\hat{\beta}_1$. By solving (5.18) with respect to $\hat{\beta}_0$ we get

$$-\sum Y_i + nb_1 + \hat{\beta}_1 \sum X_i = 0$$
$$-\frac{1}{n}\sum Y_i + \frac{1}{n}nb_1 + \frac{1}{n}\hat{\beta}_1 \sum X_i = 0$$
$$-\bar{Y} + \hat{\beta}_0 + \hat{\beta}_1 \bar{X} = 0$$
$$\hat{\beta}_0 = \bar{Y} - \hat{\beta}_1 \bar{X}$$

And by solving (5.19) with respect to $\hat{\beta}_1$, we get

$$-\sum Y_i X_i + \hat{\beta}_0 \sum X_i + \hat{\beta}_1 \sum X_i^2 = 0$$
$$-\sum Y_i X_i + (\bar{Y} - \hat{\beta}_1 \bar{X}) \sum X_i + \hat{\beta}_1 \sum X_i^2 = 0$$
$$-\hat{\beta}_1 \bar{X} \sum X_i + \hat{\beta}_1 \sum X_i^2 = \sum Y_i X_i - \bar{Y} \sum X_i$$
$$-\hat{\beta}_1 (\bar{X} \sum X_i - \sum X_i^2) = \sum Y_i X_i - \bar{Y} \sum X_i$$
$$\hat{\beta}_1 = \frac{\sum Y_i X_i - \bar{Y} \sum X_i}{\sum X_i^2 - \bar{X} \sum X_i} = \frac{\sum Y_i X_i - \bar{Y} n \bar{X}}{\sum X_i^2 - \bar{X} n \bar{X}} = \frac{\sum (Y_i - \bar{Y})(X_i - \bar{X})}{\sum (X_i - \bar{X})^2}$$

Hence, $\hat{\beta}_1 = \frac{\sum (Y_i - \bar{Y})(X_i - \bar{X})}{\sum (X_i - \bar{X})^2}$ and $\hat{\beta}_0 = \bar{Y} - \hat{\beta}_1 \bar{X}$. We can thus calculate $\hat{\beta}_1$ using the covariance between X and Y divided by the variance of X, while we find $\hat{\beta}_0$ from $\hat{\beta}_1$ and the mean of Y and X.

5.8.2 *Calculating the Numbers from a Regression Output*

Through the chapters in the book, it has been shown in several ways how regression analysis can be done both by hand and in Gretl and R. The calculations below show where most of the numbers in the output from Gretl and R have been calculated after doing a simple regression of the model.

$$wage_i = \beta_0 + \beta_1 edu_i + u_i$$

with the sample we previously used of eight people and the variables wage and education.

For the simple regression where wage is the dependent variable and education is the independent variable, we get the following output:

Model 1: OLS, using observations 1–8
Dependent variable: wage

	Coefficient	Std. Error	*t*-ratio	p-value
const	112.308	48.2979	2.325	0.0590*
edu	71.9231	11.0081	6.534	0.0006***

Mean dependent var	400.0000	S.D. dependent var	148.0347
Sum squared resid	18903.85	S.E. of regression	56.13057
R^2	0.876768	Adjusted R^2	0.856229
$F(1, 6)$	42.68850	P-value(F)	0.000614
Log-likelihood	−42.42222	Akaike criterion	88.84445
Schwarz criterion	89.00333	Hannan–Quinn	87.77285

Test for normality of residual –
Null hypothesis: error is normally distributed
Test statistic: $\chi^2(2) = 2.34886$
with p-value = 0.308994
`Analysis` > ANOVA gives

Analysis of Variance:

Sum of squares	df	Mean square	
Regression	134496	1	134496
Residual	18903.8	6	3150.64
Total	153400	7	21914.3

```
R^2 = 134496 / 153400 = 0,876768
F(1, 6) = 134496 / 3150.64 = 42,6885 [p-value 0.0006]
```

In the calculations below, wage is denoted by Y and education by X so that we get

$$Y_i = \beta_0 + \beta_1 X_i + u_i$$

From when we calculated the correlation coefficient in Table 3.1, we found much of what we needed to calculate the values we got after doing a simple regression analysis. We hadHence, we have

Hence, we have $n = 8$ observation pairs.

First, the means of X and Y are calculated (where the mean of Y, $\bar{Y}$ is "Mean dependent var" in the output).

$$\bar{X} = \frac{1}{n}\sum_{i=1}^{n} X_i = \frac{1}{8}(2+3+1+5+5+4+7+5) = \frac{32}{8} = 4$$

$$\bar{Y} = \frac{1}{n}\sum_{i=1}^{n} Y_i = \frac{1}{8}(320+290+170+480+510+410+650+370)$$
$$= \frac{3200}{8} = 400$$

In order to estimate the coefficients (found in the column `Coefficient`), in the model $Y_i = \beta_0 + \beta_1 x_i + u_i$ using the least squares method, we used

$$\hat{\beta}_1 = \frac{\sum_{i=1}^{n}(X_i - \bar{X})(Y_i - \bar{Y})}{\sum_{i=1}^{n}(X_i - \bar{X})^2} = \frac{1870}{26} = 71.923 \tag{5.20}$$

$$\hat{\beta}_0 = \bar{Y} - \hat{\beta}_1 \bar{X} = 400 - 71.923 \cdot 4 = 112.308 \tag{5.21}$$

Hence, we get the estimated regression line

$$\hat{Y} = 112.3 + 71.9 \cdot X_i$$

Furthermore, we can calculate the `S.D. dependent variable`, which is the standard deviation of Y:

$$s_y = \sqrt{\frac{(Y - \bar{Y})^2}{n-1}} = \sqrt{\frac{153400}{7}} = 148.03 \tag{5.22}$$

We can also calculate the following deviations:

- $\text{TSS} = \sum_{i=1}^{n}(Y_i - \bar{Y})^2$ (total sum of squares)
- $\text{RSS} = \sum_{i=1}^{n}(Y_i - \hat{Y}_i)^2 = \sum_{i=1}^{n}(Y_i - \hat{\beta}_0 - \hat{\beta}_1 x_i)^2 = \sum_{i=1}^{n} e_i^2$ (residual sum of squares)
- $\text{ESS} = \sum_{i=1}^{n}(\hat{Y}_i - \bar{Y})^2 = \sum_{i=1}^{n}(\hat{\beta}_0 + \hat{\beta}_1 x_i - \bar{Y})^2$ (explained sum of squares)

Total sum of squares is found in the table above

$$\text{TSS} = \sum_{i=1}^{n}(Y_i - \bar{Y})^2 = 153400 \tag{5.23}$$

We have to do some more calculations in order to find RSS and ESS, by first calculating $\hat{Y}$ for all observed values of the independent variable and then calculate the sum of the squared deviations.

From the sums in the final row, we get $\text{RSS} = \sum_{i=1}^{n}(Y_i - \hat{Y}_i)^2 = 18{,}903.8$ and $\text{ESS} = \sum_{i=1}^{n}(\hat{Y}_i - \bar{Y})^2 = 134{,}496.2$, in line with the ANOVA output.

R-squared can then be calculated as

$$R^2 = \frac{\text{ESS}}{\text{TSS}} = \frac{134496}{153400} = 0.8768 \tag{5.24}$$

We can also calculate `S.E. of regression`

$$\hat{\sigma} = \sqrt{\hat{\sigma}} = \sqrt{\frac{\text{RSS}}{n-2}} = \sqrt{\frac{18903.85}{8-2}} = \sqrt{3150.64} = 56.13$$

Table 5.4 Predicted dependent variable and error terms

i	X	Y	$\hat{Y}$	$\hat{Y}-\bar{Y}$	$(\hat{Y}-\bar{Y})^2$	$Y-\hat{Y}$	$(Y-\hat{Y})^2$	X^2
1	2	320	256.2	−143.8	20,691.7	63.8	4,076.3	4
2	3	290	328.1	−71.9	5,172.9	−38.1	1,449.9	9
3	1	170	184.2	−215.8	46,556.4	−14.2	202.5	1
4	5	480	471.9	71.9	5,172.9	8.1	65.2	25
5	5	510	471.9	71.9	5,172.9	38.1	1,449.9	25
6	4	410	400.0	0.0	0.0	10.0	100.0	16
7	7	650	615.8	215.8	46,556.4	34.2	1,171.7	49
8	5	370	471.9	71.9	5,172.9	−101.9	10,388.3	25
SUM	32	3,200	3,200.0	0.0	134,496.2	0.0	18,903.8	154

Adjusted R-squared (where k is the number of slope coefficients, here: 1; β_1) is

$$\bar{R}^2 = 1 - \frac{\text{SSE}/(n-k-1)}{SST/(n-1)} = 1 - \frac{18904/(8-2)}{153400/(8-1)} = 0.856$$

The standard error of the estimated coefficients `Std. Error` is

$$\text{SE}(\hat{\beta}_1) = \sqrt{\frac{\sigma^2}{\sum(x_i-\bar{x})^2}} = \sqrt{\frac{3150.64}{26}} = 11.01$$

$$\text{SE}(\hat{\beta}_0) = \sqrt{\frac{\sigma^2\sum x_i^2}{n\sum(x_i-\bar{x})^2}} = \sqrt{\frac{3150.64\cdot 154}{8\cdot 26}} = 48.298$$

The next column in the output, `t-ratio`, is the test statistic for the coefficient with the null hypothesis that it is equal to zero

$$T = \frac{\hat{\beta}_1}{\text{SE}(\hat{\beta}_1)} = \frac{71.92}{11.01} = 6.53$$

$$T = \frac{\hat{\beta}_0}{\text{SE}(\hat{\beta}_0)} = \frac{112.308}{48.2979} = 2.325$$

`p-value` is the associated p-value that provides the probability of getting a test observer that is at least as extreme as the one we have observed, given that the null hypothesis is true. This can, e.g., be found if you have a detailed t-distribution table or through Gretl or R.

In the table `Analysis of Variance`, the first column shows ESS (explained sum of squares/«Regression»), RSS (residual sum of squares), and TSS (total sum of squares). The next column shows the corresponding number of degrees of freedom (df). These are $k = 1$ for ESS, $n-k-1 = 8-1-1 = 6$

for RSS and $n - 1 = 8 - 1 = 7$ for RSS, where k is the number of slope coefficients (1: β_1) and n is number of observations (here: 8). `R^2` shows R-squared and that this is calculated as ESS/RSS.

`F(1,6)` (which is both in the table from the regression and the ANOVA table) is the test statistic to test the multiple hypotheses (more on this in the next chapter) $H_0 : \beta_1 = \beta_2 = \cdots = \beta_{p-1} = 0$ against $H_A : \beta_j \neq 0$ for at least one value of j. 1 is degrees of freedom in the numerator and 6 degrees of freedom in the denominator in this case. We then get the test statistic

$$
\begin{aligned}
F &= \frac{(\text{RSS}_{\text{null}} - \text{RSS})/m}{\text{RSS}/(n-k-1)} = \frac{(\text{TSS} - \text{RSS})/m}{\text{RSS}/(n-k-1)} \\
&= \frac{(153400 - 18903.8)/1}{18903.8/6} = \frac{134496.2}{3150.6} = 42.69
\end{aligned}
$$

The p-value `P-value(F)=0.0006` can be found if, for example, has a detailed table for the F-distribution. More about F-tests in Sect. 6.4.

As we will look at in more detail in Sect. 6.4, we can calculate the test statistic for an F-test by finding the RSS for the model under the null hypothesis, where we get the model $Y_i = \beta_0 + u_i$. This model will have RSS (sum squared deviation between predicted and observed Y_i) equal to TSS (sum squared deviation between observed Y_i and the mean of observed Y $\bar{Y}_i$), $\text{RSS}_{null} = \text{TSS}$.

Reference

Wooldridge, J. (2019). *Introductory econometrics: A modern approach.* Cengage Learning.

CHAPTER 6

Multiple Regression

In this chapter, we extend simple linear regression to include more than one explanatory variable. This alters the way we interpret our estimated effects, even if the procedure still follows the ordinary least squares method. Related to this is the problem of heteroskedasticity, instrumental variables and omitted variable bias, which is also covered.

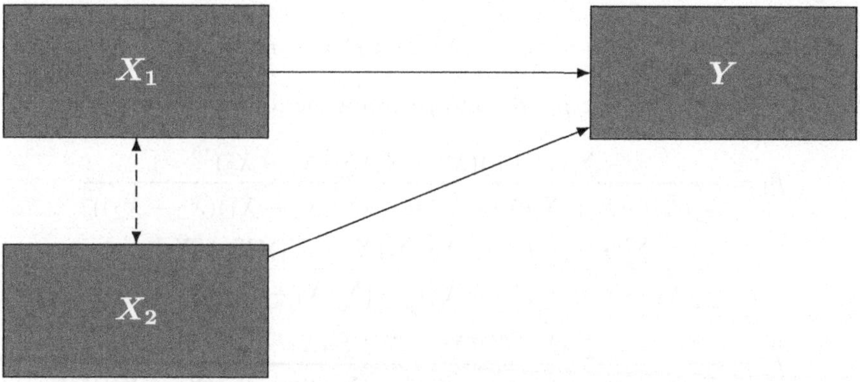

Instead of having a dependent variable (Y) and one explanatory variable (X), we can have multiple explanatory variables. This results in a **multiple**

Supplementary Information The online version contains supplementary material available at https://doi.org/10.1007/978-3-031-53142-2_6.

B. Karlsen Kivedal, *Applied Statistics and Econometrics*, https://doi.org/10.1007/978-3-031-53142-2_6

regression model:

$$Y_i = \beta_0 + \beta_1 X_{1i} + \beta_2 X_{2i} + \cdots + \beta_k X_{ki} + u_i \tag{6.1}$$

In this model, there are k explanatory variables ($X_1, \ldots, X_k$) and k+1 coefficients ($\beta_0, \beta_1, \ldots, \beta_k$). In the simple regression model from the previous chapter, there was one explanatory variable, X, resulting in k = 1 and k + 1 = 2 coefficients (β_0 and β_1). Therefore, we are now estimating a model with an intercept as before (β_0) and k different slopes that indicate how an explanatory variable can influence the dependent variable.

Often, we want to measure the effects of various variables on a dependent variable. Consequently, we may have multiple explanatory variables in a model instead of creating a separate model for each explanatory variable. This approach allows us to estimate effects while controlling for other variables. Therefore, multiple regression is used to estimate the effect of several different variables on a dependent variable simultaneously, with the estimated effects assuming that all the other variables, known as control variables, remain constant. This is necessary to calculate the isolated effect of each explanatory variable.

6.1 Manual Calculation

A regression model with two explanatory variables can be written as follows:

$$Y_i = \beta_0 + \beta_1 X_{1,i} + \beta_2 X_{2,i} + u_i \tag{6.2}$$

We can manually estimate β_0, β_1, and β_2 using the following formulas:

$$\hat{\beta}_1 = \frac{\sum(Y-\bar{Y})(X_1-\bar{X}_1)\sum(X_2-\bar{X}_2)^2}{\sum(X_1-\bar{X}_1)^2\sum(X_2-\bar{X}_2)^2 - \left(\sum(X_1-\bar{X}_1)(X_2-\bar{X}_2)\right)^2} - \frac{\sum(Y-\bar{Y})(X_2-\bar{X}_2)\sum(X_1-\bar{X}_1)(X_2-\bar{X}_2)}{\sum(X_1-\bar{X}_1)^2\sum(X_2-\bar{X}_2)^2 - \left(\sum(X_1-\bar{X}_1)(X_2-\bar{X}_2)\right)^2}$$

$$\hat{\beta}_2 = \frac{\sum(Y-\bar{Y})(X_2-\bar{X}_2)\sum(X_1-\bar{X}_1)^2}{\sum(X_1-\bar{X}_1)^2\sum(X_2-\bar{X}_2)^2 - \left(\sum(X_1-\bar{X}_1)(X_2-\bar{X}_2)\right)^2} - \frac{\sum(Y-\bar{Y})(X_1-\bar{X}_1)\sum(X_1-\bar{X}_1)(X_2-\bar{X}_2)}{\sum(X_1-\bar{X}_1)^2\sum(X_2-\bar{X}_2)^2 - \left(\sum(X_1-\bar{X}_1)(X_2-\bar{X}_2)\right)^2}$$

$$\hat{\beta}_0 = \bar{Y} - \hat{\beta}_1\bar{X}_1 + \hat{\beta}_2\bar{X}_2$$

We then get the estimated model:

$$\hat{Y}_i = \hat{\beta}_0 + \hat{\beta}_1 X_{1,i} + \hat{\beta}_2 X_{2,i} \tag{6.3}$$

Hence, we do not have a straight line as with simple regression, but a **linear function** that describes the relationship between three variables: Y, X_1, and X_2.

Calculating these values manually can be time-consuming because it involves numerous products between different sums of squared or multiplied deviations. To construct a table for these calculations would be significantly larger than what we needed in simple regression, where we only needed $\sum(X_i - \bar{X})^2$ and $\sum(X_i - \bar{X})(X_i - \bar{Y})$, as we computed in Tables 3.1 and 5.1.

The key insight from these formulas is that both the covariance between the explanatory variables and the correlation between each explanatory variable and the dependent variable has an impact on the estimated coefficients ($\sum(X_1 - \bar{X}_1)(X_2 - \bar{X}_2)$ is $(n - 1)$ times the covariance between X_1 and X_2). Each estimated effect must thus be considered in the context of the other explanatory variables (or in this case, the other explanatory variable of the two). This is why we always need to interpret the estimated coefficients as measuring the effect of one variable on the dependent variable while holding the other variables constant. We often refer to this as isolating the effect of each variable on the dependent variable, measuring the **isolated effect**, and taking into account the other explanatory variables as **control variables**.

6.2 Interpreting Estimated Parameters in Multiple Regression

In simple regression, we calculated the slope as $\hat{\beta}_1 = \sum(X_i - \bar{X})(Y_i - \bar{Y})/\sum(X_i - \bar{X})^2$. In this formula, the slope represents the relationship between how X and Y vary together and the sum of squared deviations for X. As we saw in the formula for $\hat{\beta}_1$ for multiple regression with two explanatory variables above, it also incorporates the variation of the second explanatory variable X_2 and how the second explanatory variable varies with X_1 and Y. This means that when estimating a slope in a multiple regression model, the estimated value will tell us how the explanatory variable affects the dependent variable while considering the variation in the other explanatory variables. This mathematical relationship can be illustrated through what we know as the calculation of the **partial derivative**.

6.2.1 *Partial Derivatives (Mathematics)*

Let us assume the following simple mathematical model:

$$Y = 15 + 3X + 17Z$$

If we differentiate Y with respect to X, we get $\frac{\partial Y}{\partial X} = 3$ **ceteris paribus**, which means "all else being equal", and $\frac{\partial Y}{\partial Z} = 17$ ceteris paribus (all else being equal).

All else being equal means that we are **controlling** for other factors. Since the partial derivative of Y with respect to X is 3 ceteris paribus, it implies that the effect is measured when we **hold Z constant** (controlling for changes in Z). $\frac{\partial Y}{\partial Z} = 17$ ceteris paribus means that if Z increases by 1, Y increases by 17, given that X does not change (controlling for changes in X).

6.2.2 Interpretation If We Have Two Explanatory Variables

The main difference in estimating the two simple regression models:

$$Y_i = \beta_0 + \beta_1' X_{1i} + u_i \tag{6.4}$$

and

$$Y_i = \beta_0 + \beta_2' X_{2i} + u_i \tag{6.5}$$

or estimating the multiple regression model:

$$Y_i = \beta_0 + \beta_1 X_{1i} + \beta_2 X_{2i} + u_i \tag{6.6}$$

is that we control for the other variable when estimating the coefficients in the multiple regression model.

In the simple regression models (6.4) and (6.5), $\hat{\beta}_1'$ and $\hat{\beta}_2'$ measure the effect of X_1 and X_2 on Y_i, respectively. In the multiple regression model (6.6), the definition of the estimated parameters is slightly different: $\hat{\beta}_1$ measures the effect of X_1 on Y_i, controlling for X_2, and $\hat{\beta}_2$ measures the effect of X_2 on Y_i, controlling for X_1. Controlling for the other variable means that the estimated effect refers to the effect given that the variable we control for is held constant. This is often referred to as "ceteris paribus", which is Latin and can be translated to "all else being equal", meaning that the other variables are held constant (unchanged).

In the model

$$Y_i = \beta_0 + \beta_1 X_{1,i} + \beta_2 X_{2,i} + u_i$$

which, in estimated form, becomes a multiple regression model with two explanatory variables

$$\hat{Y}_i = \hat{\beta}_0 + \hat{\beta}_1 X_{1,i} + \hat{\beta}_2 X_{2,i}$$

we have

$$\frac{\partial \hat{Y}}{\partial X_1} = \hat{\beta}_1$$

$$\frac{\partial \hat{Y}}{\partial X_2} = \hat{\beta}_2$$

Table 6.1 Dataset: wage, education, and experience

i	*1*	*2*	*3*	*4*	*5*	*6*	*7*	*8*
wage	2	3	1	5	5	4	7	5
edu	320	290	170	480	510	410	650	370
expr	5	3	1	12	40	25	35	4

Just like in simple regression, the estimated coefficients $\hat{\beta}_1$ and $\hat{\beta}_2$ show the effect on the dependent variable Y if X_1 and X_2 increase by 1, respectively. However, now the effect will also depend on the other explanatory variable (or others if we have more than two explanatory variables) being held constant (ceteris paribus, all else being equal). In other words, we control for the other explanatory variables.

Interpreting slope coefficients in multiple regression

The estimated coefficient $\hat{\beta}_i$ represents the change in the dependent variable if the explanatory variable (associated with $\hat{\beta}_i$) increases by 1, provided that the other explanatory variables remain unchanged. This is also known as "holding everything else constant" (ceteris paribus) or controlling for the other explanatory variables.

6.3 Example (in Gretl and R)

We will now expand the dataset of the eight individuals to include a variable indicating the number of years of relevant experience they had before starting their current job.

To perform multiple linear regression in Gretl, we follow the same procedure as in simple linear regression. We choose `Model > Ordinary Least Squares` (or click the button marked $\hat{\beta}$ on the lower toolbar). Then, we select the dependent variable, and we can choose multiple explanatory variables (regressors). For example, in the dataset containing wage, education, and experience (Fig. 6.1).

This gives

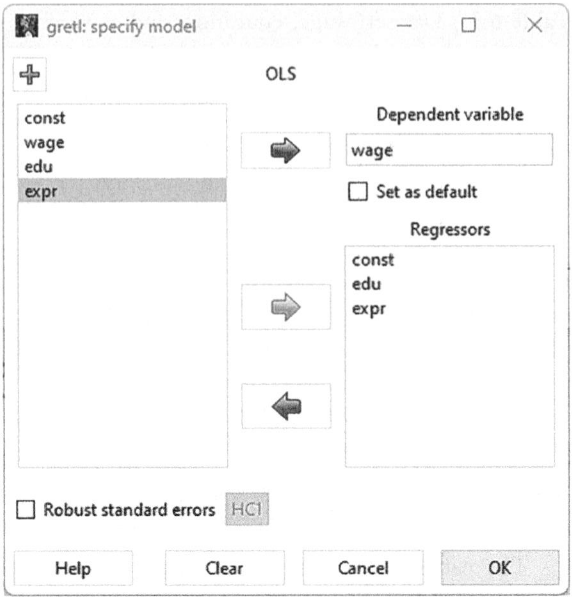

Fig. 6.1 Multiple regression, variable selection

Model 2: OLS, using observations 1–8
Dependent variable: wage

	Coefficient	Std. Error	t-ratio	p-value
const	133.880	39.1718	3.418	0.0189**
edu	53.9200	11.9554	4.510	0.0063***
expr	3.22813	1.48206	2.178	0.0813*

Mean dependent var	400.0000	S.D. dependent var	148.0347
Sum squared resid	9699.947	S.E. of regression	44.04531
R^2	0.936767	Adjusted R^2	0.911474
$F(2, 5)$	37.03630	p-value(F)	0.001005
Log-likelihood	−39.75324	Akaike criterion	85.50649
Schwarz criterion	85.74481	Hannan–Quinn	83.89909

The output is quite similar to what we obtained in simple linear regression in the previous chapter, but now there is an additional row for the variable "expr". Note that the coefficients now measure the effect of a one-unit change in the explanatory variable on the response variable, all else being equal. This means we control for the other explanatory variables. In this case, it means that we control for education when interpreting the coefficient for experience and control for experience when interpreting the coefficient for education.

For comparison, consider the following effects from the example:

- In the simple regression model $wage_i = \beta_0 + \beta_1 edu + u_i$, we estimated $\hat{\beta}_1 = 71.9$.
  - If a person has one more year of education, we expect their wage to be 71,900 higher.
- In the multiple regression model $wage_i = \beta_0 + \beta_1 edu + \beta_2 exper + u_i$, we estimated $\hat{\beta}_1 = 53.9$.
  - If a person has one more year of education, we expect their wage to be 53,900 higher, *given that their experience is the same.*

In the multiple regression model, we are comparing two people with a one-year difference in education, assuming they have the same amount of experience. In the simple regression model, we compare two people with a one-year difference in education, but we do not know whether they have different amounts of experience. Since high education is correlated with high experience in our sample (correlation of 0.69, as seen in the Gretl output below), much of the effect we found in the simple regression model also describes the effect of differences in experience.

```
Correlation Coefficients, using the observations 1 - 8
5% critical value (two-tailed) = 0.7067 for n = 8

          wage        edu        expr
        1.0000     0.9364     0.8243  wage
                   1.0000     0.6913  edu
                              1.0000  expr
```

As with simple regression, you can perform multiple regression using the command-line interface in Gretl. You can access the command-line interface by pressing Ctrl+S or clicking on the "Open Gretl console" button on the toolbar at the bottom of the main window. Then, you can enter the following command:

Gretl code 6.1: Multiple regression in Gretl

```
ols wage const edu expr
```

Additional explanatory variables can be added by including them further in the command line after the dependent variable and constant term. The order of the explanatory variables doesn't affect the results but will only determine the order of rows in the Gretl output. See 6.3 for reference.

In R, we can use the lm function and add more variables to perform multiple regression. We can add variables by adding it after a plus sign between the explanatory variables in the code: `lm(wage ~ edu + expr)`.

R code 6.1: Multiple regression in R

```
multiplereg <- lm(wage ~ edu + expr)
summary(multiplereg)
```

```
Call:
lm(formula = wage ~ edu + expr)

Residuals:
1      2      3      4      5      6      7      8
62.14 -15.32 -21.03  37.78 -22.61 -20.26  25.69 -46.39

Coefficients:
Estimate Std. Error t value Pr(>|t|)
(Intercept)  133.880     39.172   3.418  0.01888 *
edu           53.920     11.955   4.510  0.00634 **
expr           3.228      1.482   2.178  0.08130 .
---
Signif. codes:
0 '***' 0.001 '**' 0.01 '*' 0.05 '.' 0.1 ' ' 1

Residual standard error: 44.05 on 5 degrees of freedom
Multiple R-squared:  0.9368,Adjusted R-squared:  0.9115
F-statistic: 37.04 on 2 and 5 DF,  p-value: 0.001005
```

6.3.1 *Example: Used Car Price and Fuel Consumption*

To further illustrate the implications of including multiple explanatory variables in a model, we will here consider another example. We will start with a simple regression model and then add additional variables. The interpretation of the explanatory variable in the simple regression model will change, as the estimated slope coefficients in a multiple regression model also control for the other explanatory variables in the model.

We have data for 20 used cars that were sold, and for each car, we know the price it was sold for in euros (price_i) and the fuel consumption measured in the number of liters of gasoline per 1000 kilometers (gasuse_i).

To understand how fuel consumption affects the price, we can estimate the model:

$$\text{price}_i = \beta_0 + \beta_1 \text{gasuse}_i + u_i$$

We then get

Model 1: OLS, using observations 1–20
Dependent variable: price

	Coefficient	Std. Error	*t*-ratio	*p*-value
const	−1825.33	1800.72	−1.014	0.3242
gasuse	75.4096	24.2443	3.110	0.0060***

Mean dependent var	3713.500	S.D. dependent var	1443.837
Sum squared resid	25,762,065	S.E. of regression	1196.339
R^2	0.349585	Adjusted R^2	0.313451
$F(1, 18)$	9.674636	*p*-value(F)	0.006040
Log-likelihood	−169.0656	Akaike criterion	342.1312
Schwarz criterion	344.1226	Hannan–Quinn	342.5199

If gasoline consumption increase by 1 liter per 1000 km, the price is expected to increase by 75.41 euros.

We also know the engine power of each car, measured in the number of horsepower (hp_i). To account for engine power, we can estimate the model:

$$\text{price}_i = \beta_0 + \beta_1 \text{gasuse}_i + \beta_2 hp_i + u_i$$

We then get

Model 2: OLS, using observations 1–20
Dependent variable: price

	Coefficient	Std. Error	t-ratio	p-value
const	−2249.54	1515.85	−1.484	0.1561
gasuse	−12.4189	36.1412	−0.3436	0.7353
hp	51.8884	17.6591	2.938	0.0092***

Mean dependent var	3713.500	S.D. dependent var	1443.837
Sum squared resid	17,085,056	S.E. of regression	1002.499
R^2	0.568653	Adjusted R^2	0.517907
$F(2, 17)$	11.20573	p-value(F)	0.000787
Log-likelihood	−164.9586	Akaike criterion	335.9172
Schwarz criterion	338.9044	Hannan–Quinn	336.5003

Now we can see that consumption has no significant effect on price, but the number of horsepower has a significant positive effect. Hence, when controlling for the number of horsepower, gasoline consumption has no effect on price. This indicates that engine power may have a more significant impact on price than fuel consumption. The number of horsepower may be an underlying cause of higher consumption, but engine size has a more substantial influence on price than consumption. The size of the car may also play a role, as larger cars may often have larger engines and are more expensive as a result.

We also know if a car has four wheel drive or not (this is a dummy variable since it only takes two possible values, zero and one more about this in Chapter 7). We can estimate the model:

$$\text{price}_i = \beta_0 + \beta_1 \text{gasuse}_i + \beta_2 hp_i + \beta_3 \text{fourw}_i + u_i \qquad (6.7)$$

Where the variable fourw_i is 1 if car i has four wheel drive, and 0 otherwise. We then get

Model 3: OLS, using observations 1–20
Dependent variable: price

	Coefficient	Std. Error	t-ratio	p-value
const	−1696.79	1229.35	−1.380	0.1865
gasuse	−26.3095	29.3422	−0.8966	0.3832
hp	49.0822	14.2079	3.455	0.0033***
fourw	1291.28	401.158	3.219	0.0054***

Mean dependent var	3713.500	S.D. dependent var	1443.837
Sum squared resid	10,369,837	S.E. of regression	805.0558
R^2	0.738193	Adjusted R^2	0.689104
$F(3, 16)$	15.03788	p-value(F)	0.000065
Log-likelihood	−159.9656	Akaike criterion	327.9311
Schwarz criterion	331.9141	Hannan–Quinn	328.7086

We can see that consumption still has no significant effect on the price, and furthermore, the price effect of increased horsepower has decreased from 51.9 to 49.1 (although the uncertainty with a standard error of 14.2 means that this decrease is not necessarily significant). If a car has four wheel drive, the price will be 1291 euros higher, provided that consumption and engine power remain unchanged. This effect is also significant.

In addition, the model's explanatory power has increased with the inclusion of multiple variables. The adjusted R-squared increased from 0.31 in the simple regression model to 0.52 when we included the number of horsepower, and it is 0.69 when we also include the variable indicating whether the cars are station wagons or not in the model.

Results of such models are often summarized as shown in Table 6.2[1]

Table 6.2 Results

	Dependent variable: price		
	(1)	*(2)*	*(3)*
gasuse	7.541***	−1.242	−2.631
	(2.424)	(3.614)	(2.934)
hp		5.189***	4.908***
		(1.766)	(1.421)
fourw			129.128***
			(40.116)
Constant	−182.533	−224.954	−169.679
	(180.072)	(151.585)	(122.935)
Observations	20	20	20
R^2	0.350	0.569	0.738
Adjusted R^2	0.313	0.518	0.689

*$p < 0.1$; **$p < 0.05$; ***$p < 0.01$

[1] This table was created using the "Stargazer" package in R. Refer to the R script in the online resources for the code.

6.4 F-Test

Multiple hypothesis testing involves testing hypotheses simultaneously. Up to this point, we have only tested whether $\beta_1 = 0$ or $\beta_2 = 0$ individually. The hypotheses become:

$$H_0\text{: } \beta_1 = a_1 \text{ and } \beta_2 = a_2 \text{ and, ..., and } \beta_k = a_k$$
$$H_A\text{: } \beta_1 \neq a_1 \text{ and/or } \beta_2 \neq a_2 \text{ and/or, ..., and/or } \beta_k \neq a_k$$

where a_i are constants (e.g., often 0). This means that if at least one of the values in the null hypothesis is not as specified, the null hypothesis must be rejected.

To conduct such a hypothesis test, we need to compare the model we are working with to the model as it appears if we impose the restrictions from the null hypothesis. For example, if we have the null hypothesis $\beta_2 = \beta_3 = 0$ for the model:

$$Y_i = \beta_0 + \beta_1 X_{1,i} + \beta_2 X_{2,i} + \beta_3 X_{3,i} + u_i$$

we get the model:

$$Y_i = \beta_0 + \beta_1 X_{1,i} + u_i$$

We will refer to this model as the model under the null hypothesis, and it will have RSS_{null} and R^2_{null}.

The **test statistic** can be calculated as:

$$F = \frac{(RSS_{null} - RSS)/m}{RSS/(n-k-1)} = \frac{(R^2 - R^2_{null})/m}{(1-R^2)/(n-k-1)}$$

where:

- *m*: the number of statements (number of equal signs "<<=>>") in H_0.
- *n*: the number of observations.
- *k*: the number of estimated slope coefficients in the model.
- *RSS*: the regression sum of squares in the model.
- RSS_{null}: the regression sum of squares in the model with the restrictions from the null hypothesis implemented.

The **critical value** is found in the F-distribution with *m* degrees of freedom in the numerator and $n - k - 1$ degrees of freedom in the denominator; $cv = F_\alpha(m, n-k-1)$.

6.4.1 *Example F-Test*

For the model:

$$Y_i = \beta_0 + \beta_1 X_{1,i} + \beta_2 X_{2,i} + u_i$$

we may be interested in testing whether at least one of the explanatory variables in the model has a significant effect on the dependent variable. We can do this through the hypotheses:

$$H_0\colon \beta_1 = 0 \text{ and } \beta_2 = 0$$
$$H_A\colon \beta_1 \neq 0 \text{ and/or } \beta_2 \neq 0$$

If at least one of the coefficients is different from zero, we reject the null hypothesis.

To calculate the test statistic, we need R^2 or RSS for the model (which is the estimated model $\hat{Y}_i = \hat{\beta}_0 + \hat{\beta}_1 X_{1,i} + \hat{\beta}_2 X_{2,i}$). In the model where wage was the dependent variable and education and experience were explanatory variables, we found $R^2 = 0.937$ and $RSS = 9699.95$. If we impose the restrictions from H_0, i.e., we set $\beta_1 = \beta_2 = 0$ in the model, we obtain the estimated model under the null hypothesis:

$$\hat{Y}_i = \hat{\beta}_0 + 0 \cdot X_{1,i} + 0 \cdot X_{2,i}$$
$$\hat{Y}_i = \hat{\beta}_0$$

Since this is a model for Y_i with only the intercept β_0, the model cannot explain any of the variation in Y_i. Therefore, we have $R^2_{null} = 0$. In addition, we have $ESS_{null} = 0$ (explained sum of squares in the model under the null hypothesis is zero), so $RSS_{null} = TSS - ESS_{null} = TSS - 0 = TSS$. RSS_{null} is, therefore, equal to the total sum of squares for Y_i that we previously found to be $TSS = \sum(Y_i - \bar{Y})^2 = 153{,}400$.

R-squared and RSS in a model with only an intercept

In a model for a variable with only an intercept and no explanatory variables, R-squared is zero, and RSS is equal to the TSS of the dependent variable.

The test statistic is therefore:

$$F = \frac{(R^2 - R^2_{null})/m}{(1 - R^2)/(n - k - 1)} = \frac{(0.937 - 0)/2}{(1 - 0.937)/(8 - 3)} = \frac{0.469}{0.0126} = 37.22$$

So we have $m = 2$ statements in the null hypothesis and $k = 2$ estimated slope coefficients in the model (β_1 and β_2).

If we had used all decimals, we would have obtained $F = 37.03630$. We find this number again as "F(2, 5)" in the Gretl output from the multiple regression model, where "*p*-value (F)" at 0.001005 is the associated tail probability for the number 37.03630 on the F-distribution.

We can find the critical value in the F-distribution with $m = 2$ degrees of freedom in the numerator and $n - k = 8 - 3 = 5$ degrees of freedom in the

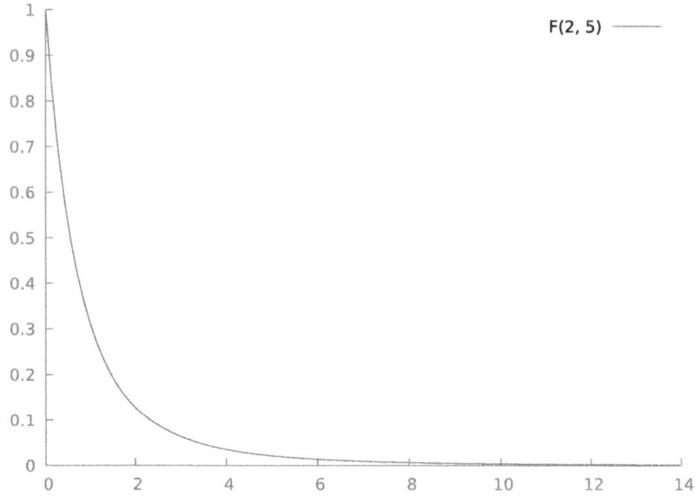

Fig. 6.2 F-distribution with two degrees of freedom in the numerator and five in the denominator; $F(2, 5)$

denominator. In Gretl, you can find this by going to `Tools > Statistical tables` and selecting the `F` tab. Enter 2 for `dfn` (Degrees of Freedom numerator) and for `dfd` (Degrees of Freedom Denominator). For the `right-tail probability`, choose the significance level, e.g., 0.05 if you have a 5% significance level. In R, you can find the critical value by typing `qf(0.95, 2, 5)`, where $1 - \alpha = 1 - 0.05 = 0.95$ is the first parameter followed by the degrees of freedom in the numerator and denominator.

You do not need to worry about using $\alpha/2$ or one-tailed and two-tailed tests in these tests, as all tests are one-sided. You get the critical value of 5.786 from the F-distribution. Since the test statistic is higher than the critical value, we reject the null hypothesis, which means we find support for the idea that education or experience, or both, affect wages. The F-distribution corresponding to our example is shown in Fig. 6.2.

Eksemplet i Gretl
After estimating the multiple regression model, we can select `Tests > Omit variables`. Here, we can choose `edu` and `expr` as the variables we want to omit, in line with the restrictions we had in the null hypothesis of the F-test ($\beta_1 = \beta_2 = 0$). After clicking `OK`, we get the results displayed below.

```
Test on Model 2:

Null hypothesis: the regression parameters are zero for
the variables edu, expr
Test statistic: F(2, 5) = 37.0363. p-value 0.00100545
Omitting variables improved 0 of 3 information criteria.

Model 2: OLS, using observations 1-8
Dependent variable:

              coefficient   std. error   t-ratio   p-value
  ---------------------------------------------------------
  const        400.000       52.3382      7.643    0.0001   ***

Mean dependent var   400.0000   S.D. dependent var   148.0347
Sum squared resid    153400.0   S.E. of regression   148.0347
R-squared            0.000000   Adjusted R-squared   0.000000
Log-likelihood      -50.79696   Akaike criterion     103.5939
Schwarz criterion    103.6734   Hannan-Quinn         103.0581
```

We get the test statistic of 37.0363 and its associated p-value. Therefore, we can conclude that we reject the null hypothesis at a significance level of at least 0.001 (0.1%).

Example in R

To conduct the F-test in R, we can use the `anova` function to compare two models. First, we need to estimate the model under the null hypothesis, $Y_i = \beta_0 + u_i$, which is the model with only the intercept. We can estimate this by having 1 (a constant) as the sole explanatory variable. After doing this, we can find the test statistic and p-value for the test using `anova(M1, M2)`, where M1 is the model we started with, and M2 is the model under the null hypothesis.

R code 6.2: F-test in R

```
H0model <- lm(wage  1) anova(multiplereg, H0model)
```

```
Analysis of Variance Table

Model 1: wage ~ edu + expr
Model 2: wage ~ 1
Res.Df     RSS Df Sum of Sq      F   Pr(>F)
1      5   9700
2      7 153400 -2   -143700 37.036 0.001005 **
---
Signif. codes:
0 '***' 0.001 '**' 0.01 '*' 0.05 '.' 0.1 ' ' 1
```

6.5 Heteroskedasticity Tests

Since heteroskedasticity implies that the variance depends on the value of the explanatory variable, we can use a regression model and a hypothesis test to check for it. One approach is to estimate the model:

$$u_i^2 = \beta_0 + \beta_1 X_{1,i} + \beta_2 X_{2,i} + \ldots + \beta_k X_{k,i} + u_i \tag{6.8}$$

where u_i^2 is the squared error term from the regression model, $X_{1,i}, \ldots, X_{k,i}$ are the k explanatory variables, and u_i is the error term of this model. We can then conduct a multiple hypothesis test:

$$H_0: \beta_1 = \beta_2 = \ldots = \beta_k = 0$$
$$H_A: \beta_1,\ \beta_2,\ \text{and/or}\ \ldots, \beta_k \neq 0$$

using an F-test. If we reject the null hypothesis, it indicates that there are heteroskedasticity issues. This test is known as the Breusch–Pagan test (Breusch & Pagan, 1979). We can perform this test manually by saving the residuals from the estimated regression model and then estimate (6.8).

Additionally, we can include squared terms and possibly cross-products in (6.8). If we have two explanatory variables, we can have models:

$$u_i^2 = \beta_0 + \beta_1 X_{1,i} + \beta_2 X_{2,i} + \beta_3 X_{1,i}^2 + \beta_4 X_{2,i}^2 + u_i \tag{6.9}$$

$$u_i^2 = \beta_0 + \beta_1 X_{1,i} + \beta_2 X_{2,i} + \beta_3 X_{1,i}^2 + \beta_4 X_{2,i}^2 + \beta_5 X_{1,i} X_{2,i} + u_i \tag{6.10}$$

which are White (1980)'s tests for heteroskedasticity, with only squared terms and with cross-products. We can also use an F-test to determine if at least one of the explanatory variables has an effect on u_i^2 as with the Breusch–Pagan test. However, this test allows for the possibility that cross-products or squared

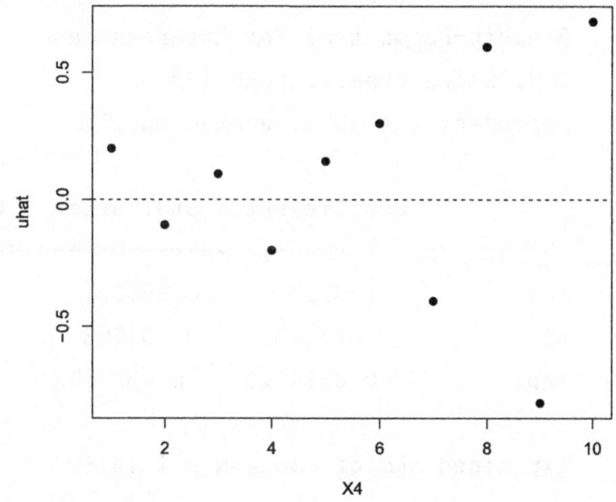

Fig. 6.3 Scatterplot between the explanatory variable X_4 and residual $\hat{u}$

terms may impact the error term, leading to the rejection of the null hypothesis of no heteroskedasticity.

If we have heteroskedasticity, we may observe this in a scatter plot between the residuals and one of the explanatory variables. Heteroskedasticity can be seen if we have a pattern that demonstrates a relationship between them, as shown in Fig. 6.3, where the residuals $\hat{u}$ are larger for high values of the explanatory variable X_4. Consequently, the model is less precise for high values of X_4 compared to low values of X_4.

6.5.1 *The Example in Gretl*

If we estimate the model:

$$\text{wage}_i = \beta_0 + \beta_1 \text{edu}_i + \beta_2 \text{expr}_i + u_i$$

and then save the residuals $\hat{u}_i$ and square them to obtain $(\hat{u}_i)^2$, we can estimate (6.8), (6.9), or (6.10), where $X_{1,i} = \text{edu}_i$ and $X_{2,i} = \text{expr}_i$ to test for heteroskedasticity.

Performing this in Gretl or R is straightforward after estimating the regression model for which we want to test for heteroskedasticity. In Gretl, you can navigate to `Tests > Heteroskedasticity`. Here, you can choose from `White's test`, `White's test (squares only)`, `Breusch-Pagan`, and `Koenker`. The last one is a variant of the Breusch–Pagan test that corrects for the possibility that residuals may not be normally distributed (Koenker, 1981).

By performing the Breusch–Pagan test, we get

```
Breusch-Pagan test for heteroskedasticity
OLS, using observations 1-8
Dependent variable: scaled uhat^2

             coefficient    std. error    t-ratio    p-value
  ----------------------------------------------------------
  const       1.19769       0.998001      1.200      0.2839
  edu         0.0732842     0.304593      0.2406     0.8194
  expr       -0.0314128     0.0377592    -0.8319     0.4434

Explained sum of squares = 1.14152

Test statistic: LM = 0.570759,
with p-value = P(Chi-square(2) > 0.570759) = 0.751729
```

Here we can see that the *p*-value is high (0.75), which means we fail to reject the null hypothesis of no heteroskedasticity. We also observe the results from the estimated model (6.8) that neither `edu` nor `expr` has a significant effect on `uhat^2`, which is the squared residuals, due to the low test statistic and high *p*-value.

6.5.2 *Example in R*

In R, we can use the `bptest()` function to conduct a Breusch–Pagan test for heteroskedasticity. However, we need to install the `lmtest` package first and then load it. You can do this as follows:

R code 6.3: Breusch-Pagan-test

```
install.packages("lmtest")
library(lmtest)
bptest(multiplereg)
bptest(multiplereg, ~ I(edu^2) + I(expr^2))
bptest(multiplereg, ~ I(edu^2) + I(expr^2) + edu*expr)
```

The first line is necessary to install the `lmtest` package, but we only need to do it once. The third line performs the Breusch–Pagan test on the `multiplereg` model, which is the regression model used as an example in this chapter. The second-to-last line shows White's test with only squared explanatory variables, while the last line shows White's test that also includes cross-products between the two explanatory variables. Performing the Breusch–Pagan test in R will provide the following output:

```
studentized Breusch-Pagan test

data:  multiplereg
BP = 1.2278, df = 2, p-value = 0.5412
```

We observe a high *p*-value, indicating that we do not have heteroskedasticity in the model.

6.5.3 Dealing with Heteroskedasticity

Since heteroskedasticity implies a relationship between unobserved errors and explanatory variables, it suggests that the model is missing something causing this bias. This could be due to the omission of a variable that accounts for this relationship.

If there is a relationship between the residuals and, for example, education, it indicates that the model is less precise for different education levels. For example, the model may be less precise for those with higher education. Adding an additional explanatory variable to account for the type of education could potentially improve model precision.

However, it is often the case that we don't have any more variables than those already included in the model from the available dataset. In such situations, one option could be to use a non-linear model. This allows the model to account for the fact that the explanatory variable has varying effects on the dependent variable at different levels of the explanatory variable. We will explore many ways to do this in Chapter 8.

Another approach is to estimate robust standard errors that account for heteroskedasticity. In Gretl, you can easily do this by checking the "Robust standard errors" option in the window where you estimate the model (see Fig. 5.5). In R, you can do this for example by installing the `estimate` package and use the `lm_robust` function; refer to the scripts in the online resources for an example. However, it's important to note that heteroskedasticity often results from a misspecified model, and it's generally better to aim for a well-specified model than resorting to the solution of robust standard errors, which can make estimates less precise (King & Roberts, 2015).

6.6 Multicollinearity

If we have a high correlation between the independent variables in a model, this may affect our results. In particular, it may increase the standard errors of the estimated coefficients. The likelihood of obtaining insignificant effects when having multicollinearity in the model is thus larger, and it is hard to conclude what the effects of each of the correlated independent variables are on the dependent variable. It is hard to deal with this, but we should be aware of the issue and look at other samples, if available, or other model specifications. This also investigates the reliability and validity of regression results.

When we included experience in the model in addition to education, which had a correlation of 0.69, the standard deviation of the parameter estimate for education increased from 11.01 to 11.96. Even if experience was significant at a 10% significance level, multicollinearity led to a more imprecise estimate of education on wages.

If we have a lot of multicollinearity in a model, we may have included too many variables and **overfitted** the model. This can be beneficial when predicting the dependent variable, but larger standard errors for the estimated coefficients make it hard to infer about the significance and importance of the explanatory variables in the model.

6.7 Instrumental Variable Regression

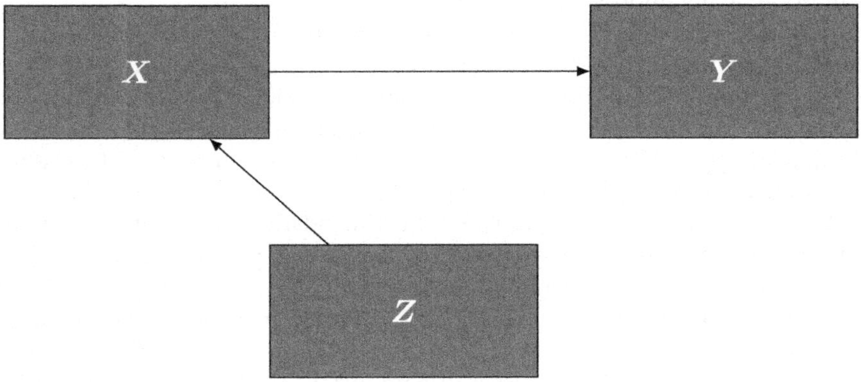

In simple linear regression, we used the example

$$Y_i = \beta_0 + \beta_1 X_i + u_i$$

where Y_i was wage and X_i education. The estimated coefficient $\hat{\beta}_1$ then measured the effect of education on wages, that is, the expected increase in wages for one additional year of education. We found that for one extra year of education, the wages were expected to increase by EUR 71.9. However, this effect on wages by education may not be the full story. We can imagine some variable that affects both wages and education. Personal ambitions or family background may have both led these persons to obtain a certain number of years of education and seeking jobs that give a certain pay.

If this is the case, it is not correct to conclude that higher education leads to higher wages. Something other than education is the actual factor affecting wages, and this factor also affects education. If this is personal ambitions and family background, we do not necessarily have a variable for this. Hence, this is part of the error term u_i in our regression model. The two variables are thus both affected by some factor outside the model, and we have a problem of **endogeneity** since the independent variable is correlated with the error term. This violates the fourth classic assumption mentioned in the previous chapter. We can deal with this problem by using an **instrumental variable**.

An instrumental variable is a variable that is correlated with the independent variable, but not with the error term. If we have an instrument Z_i, the following two conditions apply for it to be a valid instrument:

$$\text{corr}(Z_i, X_i) \neq 0$$

and

$$\text{corr}(Z_i, u_i) = 0$$

where the first condition implies that the instrument is relevant, and the second condition that it is exogenous.

The instrument therefore needs to be an observed variable. There are many potential candidates for a good instrumental variable (also often referred to as simply an instrument, a **proxy variable**, or abbreviated to IV). Card (1995) used a variable indicating whether the person grew up close to a four-year college as an instrument for educating when trying to explain wages. Hence, closeness to college may impact the amount of education for an individual, but should not be a potential independent variable to explain wages. The instrument should be related to the explanatory variable, but not to the dependent variable, except for the effect through the explanatory variable. There is no theoretical reason why distance to college when growing up should affect wages when you are an adult by itself, but it may affect the amount of education you get, and thereby have an indirect effect on wages.

In order to estimate the effect of X_i on Y_i incorporating using the instrumental variable, we use an **instrumental variable estimator**. This estimator is called the **two-stage least squares** regression (often abbreviated as **TSLS** or **2SLS**). This method involves, as the name suggests, performing an OLS regression in two steps. The first step is to estimate the effect of the instrument

Table 6.3 Dataset: wages, education, and distance to nearest college

i	*1*	*2*	*3*	*4*	*5*	*6*	*7*	*8*
wage	2	3	1	5	5	4	7	5
edu	320	290	170	480	510	410	650	370
dist	55	52	60	25	35	35	5	32

Z_i on the independent variable X_i

$$X_i = \gamma_0 + \gamma_1 Z_i + u_i \tag{6.11}$$

The fitted values from this regression, $\hat{X}_i$, are then used when estimating the effect of X_i on Y_i, thus isolating the effect of X_i on Y_i that is due to Z_i.

$$Y_i = \beta_0^{IV} + \beta_1^{IV} \hat{X}_i + u_i \tag{6.12}$$

where $\hat{\beta}_0^{IV}$ and $\hat{\beta}_1^{IV}$ are the two-stage least squares estimators.

In addition to the theoretical requirement of instrumental relevance and exogeneity which needs to be argued when including an instrument in our regression, we can also perform statistical tests for this. A commonly used test is the Hausman test, also called the Durbin–Wu–Hausman test, which tests for endogeneity of the explanatory variable(s) in the model. We will not go into details regarding this test here, but comment on it since it is provided in the regression outputs from Gretl and R.

6.7.1 Example Using an Instrumental Variable

In order to illustrate using instrumental variables in our example with wages and education for eight people, we assume that we have observed the distance to nearest college where the person grew up. Along with wages and education, we have the distance as the variable Z_i which we will use as an instrument. We therefore assume that there should be no causal effect from distance to college on wages, other than potentially affecting the degree of education. This is a **theoretical argument** for using distance to nearest college when growing up as an instrument, in order to estimate the **causal effect** of education on wages. Data for the three variables we will use are shown in Table 6.3.

The first step of the two-stage least squares regression is to estimate

$$X_i = \gamma_0 + \gamma_1 Z_i + u_i \tag{6.13}$$

This yields

Model 1: OLS, using observations 1–8
Dependent variable: edu

	Coefficient	Std. Error	t-ratio	p-value
const	7.85584	0.464810	16.90	0.0000***
dist	−0.103166	0.0113347	−9.102	0.0001***

Mean dependent var	4.000000	S.D. dependent var	1.927248
Sum squared resid	1.755913	S.E. of regression	0.540973
R^2	0.932465	Adjusted R^2	0.921209
$F(1, 6)$	82.84268	p-value(F)	0.000099
Log-likelihood	−5.285698	Akaike criterion	14.57140
Schwarz criterion	14.73028	Hannan–Quinn	13.49979

We then save $\hat{X}_i$ (the predicted values of X_i from estimating $X_i = \gamma_0 + \gamma_1 Z_i + u_i$), and use this as the explanatory variable to estimate the effect on Y_t. This provides

Model 2: OLS, using observations 1–8
Dependent variable: wage

	Coefficient	Std. Error	t-ratio	p-value
const	103.319	51.1876	2.018	0.0901*
eduhat	74.1702	11.7338	6.321	0.0007***

Mean dependent var	400.0000	S.D. dependent var	148.0347
Sum squared resid	20,027.93	S.E. of regression	57.77533
R^2	0.869440	Adjusted R^2	0.847680
$F(1, 6)$	39.95582	p-value(F)	0.000732
Log-likelihood	−42.65327	Akaike criterion	89.30655
Schwarz criterion	89.46543	Hannan–Quinn	88.23495

The estimated coefficient $\hat{\beta}_1^{IV} = 74.17$ shows that one year of additional education is associated with a wage that is 74.17 higher. This is a slightly higher estimate than what we found when using simple regression with education as the only explanatory variable. The interpretation is also different, since we now measure the effect of education explained by the distance to nearest college on wages. The independent variable in (6.12) is the predicted years of education based on college proximity when growing up. Factors that may be common for wages and education such as family background and personal abilities are thus controlled for, and we seek to measure the causal effect of education on wages.

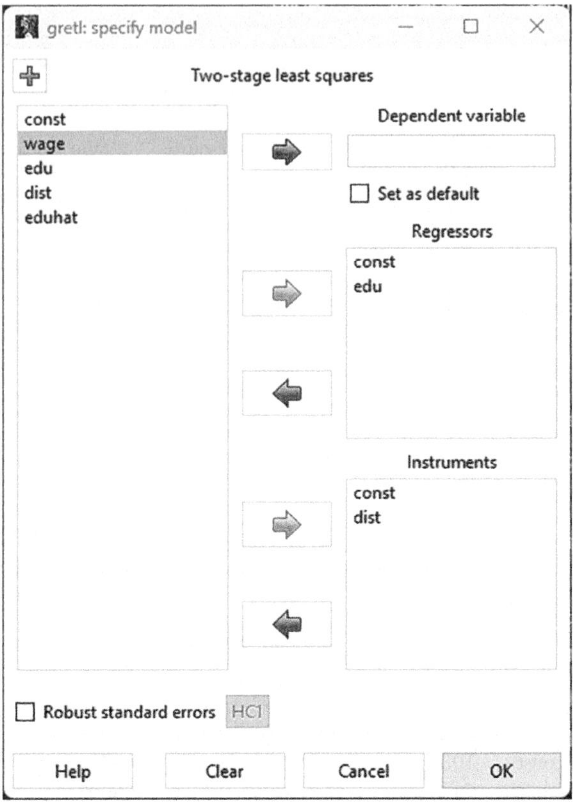

Fig. 6.4 Two-stage least squares in Gretl

Two-Stage Least Squares in Gretl

In Gretl, we can perform a two-stage least squares regression manually, as shown above, or it is possible to use the built-in function for this. This model is found by going to `Model > Instrumental variables > Two-stage least squares`. We then get the window shown in Fig. 6.4, where we select `wage` as the dependent variable, `const` and `edu` as regressors, and `const` and `dist` as instruments.

We then get

Model 3: TSLS, using observations 1–8
Dependent variable: wage
Instrumented: edu
Instruments: const dist

	Coefficient	Std. Error	t-ratio	p-value
const	103.319	49.9028	2.070	0.0838*
edu	74.1702	11.4393	6.484	0.0006***

Mean dependent var	400.0000	S.D. dependent var	148.0347
Sum squared resid	19,035.14	S.E. of regression	56.32515
R^2	0.876768	Adjusted R^2	0.856229
$F(1, 6)$	42.03975	p-value(F)	0.000640
Log-likelihood	−48.58872	Akaike criterion	101.1774
Schwarz criterion	101.3363	Hannan–Quinn	100.1058

Hausman test—
Null hypothesis: OLS estimates are consistent
Asymptotic test statistic: $\chi^2(1) = 0.848505$
with p-value = 0.356975
Weak instrument test—
First-stage $F(1, 6) = 82.8427$
Critical values for desired TSLS maximal size, when running
tests at a nominal 5% significance level:
size 10% 15% 20% 25%
value 16.38 8.96 6.66 5.53
Maximal size is probably less than 10%

We can also use code to do a 2SLS estimation in Gretl, through the command `tsls`.

Gretl code 6.2: Two-stage least squares in Gretl

```
tsls wage const edu; const dist
```

Two-Stage Least Squares in R

In R, we can estimate using two-stage least squares manually by first estimating (6.11), store the fitted values, and then estimate (6.12). It is also possible to use the function ivreg from the library AER.

R code 6.4: Two stage least squares in R

```
dist <- c(55,52,60,25,35,35,5,32)

onesls <- lm(edu ~ dist)
summary(onesls)
eduhat <- fitted(onesls)

twosls <- lm(wage ~ eduhat)
summary(twosls)

install.packages("AER") #Only necessary the first time
library(AER)
ivreg <- ivreg(wage ~ edu | dist)
summary(ivreg, diagnostics = TRUE)
```

This yields

```
Call:
ivreg(formula = wage ~ edu | dist)

Residuals:
Min          1Q    Median        3Q      Max
-104.170  -14.574    7.915    29.574   68.340

Coefficients:
Estimate Std. Error t value Pr(>|t|)
(Intercept)    103.32       49.90    2.070  0.08384 .
edu             74.17       11.44    6.484  0.00064 ***

Diagnostic tests:
df1 df2 statistic  p-value
Weak instruments    1   6      82.84 9.88e-05 ***
Wu-Hausman          1   5       0.53    0.499
Sargan              0  NA         NA       NA
---
Signif. codes:
0 '***' 0.001 '**' 0.01 '*' 0.05 '.' 0.1 ' ' 1

Residual standard error: 56.33 on 6 degrees of freedom
Multiple R-Squared: 0.8759,Adjusted R-squared: 0.8552
Wald test: 42.04 on 1 and 6 DF,  p-value: 0.0006397
```

6.8 Omitted Variable Bias

It may very well be the case that we have failed to include a relevant variable of model. This can cause **unbiased estimates** or **unbiasedness**, which violates one of the assumptions behind ordinary least squares estimation.

In the model[2]

$$\text{wage}_i = \beta_0 + \beta_1 \text{edu}_i + u_i$$

[2] This could also be illustrated in the multiple regression model where we also included experience as an explanatory, but we use the simple regression model here in order to more easily illustrate the issue.

there may have been some factors that we did not include in the model, but which actually is relevant in order to explain wages. These factors may be related both to the level of education and to wages.

Family background could be a factor that influences the degree of education. It could also be that family background affects wages through the kind of jobs the person is likely to apply for. In the single regression model with only education as the explanatory variable, all other explanatory factors are captured by the error term u_i.

Hence, we have **omitted vraiable bias** if the omitted variable is correlated with the included explanatory variable and if can determine the dependent variable. It may also be that we cannot observe the omitted variable, but it will still lead to an unbiased estimate.

If the true model is

$$\text{wage}_i = \beta_0 + \beta_1 \text{edu}_i + \beta_2 \text{family}_i + u_{1i}$$

where family_i represents family background, but we estimate the model (as in the previous chapter)

$$\text{wage}_i = \beta_0 + \beta_1 \text{edu}_i + u_{2i}$$

we see that this means that the residual u_{2i} now also contains family background which is an omitted variable. Hence, we have that $u_{2i} = \beta_2 \text{family}_i + u_{1i}$. This means that the estimate of β_1 can be biased. If there is correlation between the omitted variable and the explanatory variable, and the omitted variable affects the dependent variable, we have omitted variable bias. The OLS assumption that the error term should be uncorrelated with the explanatory variable is thus violated if we have omitted variable bias. In other words, $E(u_i|X_i) \neq 0$ if we have omitted variable bias. There is no known test for omitted variable bias, we thus needs to be careful that we have indluded all of the relevant variables in a model in order to avoid this.

References

Breusch, T. S., & Pagan, A. R. (1979). A simple test for heteroscedasticity and random coefficient variation. *Econometrica: Journal of the Econometric Society, 47*(5), 1287–1294.

Card, D. (1995). Using geographic variation in college proximity to estimate the return to schooling. In L. N. Christophides, E. K. Grant, & R. Swidinsky (Eds.), *Aspects of labour market behavior: Essays in honor of John Vanderkamp* (pp. 201–222). Toronto: University of Toronto Press.

King, G., & Roberts, M. E. (2015). How robust standard errors expose methodological problems they do not fix, and what to do about it. *Political Analysis, 23*(2), 159–179.

Koenker, R. (1981). A note on studentizing a test for heteroscedasticity. *Journal of Econometrics, 17*(1), 107–112.

White, H. (1980). A heteroskedasticity-consistent covariance matrix estimator and a direct test for heteroskedasticity. *Econometrica: Journal of the Econometric Society, 48*(4), 817–838.

CHAPTER 7

Regression Using Dummy Variables

Dummy variables are variables that can only have two different values. This provides a lot of interesting regression models. In addition to utilizing these as fixed effects and interaction effects, we also look at probability models such as probit and logit models which is relevant to use when the dependent variable is a dummy variable.

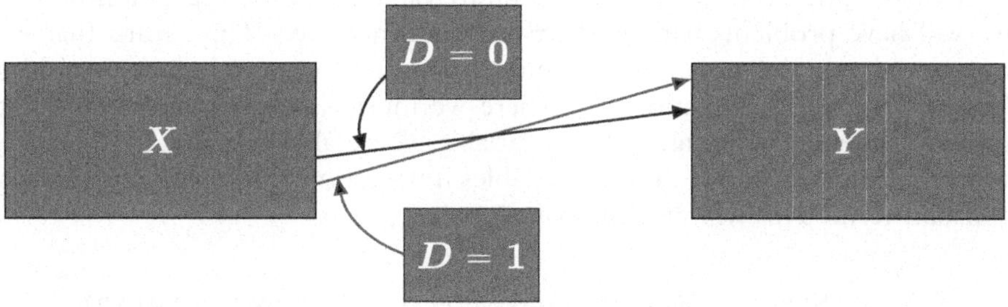

Dummy Variables (also often called binary variables or dichotomous variables) are variables that have only two possible values. For example, yes/no responses to a question, whether a car has a tow hitch or not, or whether a person is unemployed or not. We usually assign the value 1 or 0 to these dummy variables. If this variable is a dependent variable (on the left side) in a regression model, we should often use a method other than linear regression (more on this in Sects. 7.6.2 and 10.2). If the dummy variable is one of several explanatory variables, we can interpret the estimated coefficient in the

Supplementary Information The online version contains supplementary material available at https://doi.org/10.1007/978-3-031-53142-2_7.

B. Karlsen Kivedal, *Applied Statistics and Econometrics*,
https://doi.org/10.1007/978-3-031-53142-2_7

Table 7.1 Wage, education, experience, sector, and gender

i	*Wage*	*Edu*	*Expr*	*Public*	*Private*
1	320	2	5	1	0
2	290	3	3	1	0
3	170	1	1	1	0
4	480	5	12	0	1
5	510	5	40	0	1
6	410	4	25	0	1
7	650	7	35	0	1
8	370	5	4	1	0

same way as before in simple and multiple regression, but there are also several other ways to make use of dummy variables.

For example, we can have observations about the variables "public" and "private", which are 1 if the person is employed in the public or private sector, and 0 otherwise. This will give us two new variables that only have the values 0 or 1, as shown in Table 7.1.

From the data, we can see that person $i = 1$ is employed in the public sector ($\text{public}_1 = 1$, $\text{private}_1 = 0$), while person $i = 5$ is employed in the private sector ($\text{public}_5 = 0$, $\text{private}_5 = 1$).

If we include both of these two dummy variables in the regression model, we will have problems with **perfect multicollinearity**.[1] This means that we cannot estimate the model. Perfect multicollinearity occurs when two variables have a perfect linear relationship, where we must also consider the constant term, which can be thought of as a variable where all values are 1. However, we can include all of our dummy variables if we estimate the model without a constant term. This will give the same results.

7.1 Simple Regression with a Dummy Variable

If we include only one dummy variable as an explanatory variable in the model, such as the dummy variable public $_i$, we have the following regression model:

$$wage_i = \beta_0 + \beta_1 \text{public}_i + u_i$$

In this case, we obtain the following results:

[1] Multicollinearity (not perfect) means that we have a strong linear relationship between two variables, e.g., because there is a high correlation between them (close to -1 or 1). This can also create challenges, but it will still be possible to estimate the model. We saw this in the examples of multiple regression with correlations between explanatory variables. See Sect. 6.6 for more about this.

Model 1: OLS, using observations 1–8
Dependent variable: wage

	Coefficient		Std. error		t-ratio		p-value	
const	512	500	46	6146	10	99	0	0000***
public	−225	000	65	9229	−3	413	0	0143**

Mean dependent var	400.0000	S.D. dependent var	148.0347
Sum squared resid	52150.00	S.E. of regression	93.22911
R^2	0.660039	Adjusted R^2	0.603379
$F(1, 6)$	11.64909	P-value(F)	0.014265
Log-likelihood	−46.48126	Akaike criterion	96.96252
Schwarz criterion	97.12140	Hannan–Quinn	95.89092

We have estimated $\hat{\beta}_0 = 512.5$ and $\hat{\beta}_1 = -225.0$, resulting in the model $\hat{wage}_i = 512.5 - 225.0\text{public}_i$. We can also find these values by calculating the mean wages for individuals employed in the public and private sectors separately. Since individuals 1, 2, 3, and 8 are employed in the public sector ($\text{public}_i = 1$ for $i = [1, 2, 3, 8]$), and individuals 4 to 7 are employed in the private sector ($\text{public}_i = 0$ for $i = [4, 5, 6, 7]$), we obtain:

$$\overline{wage}|(\text{public}_i = 0) = \frac{480 + 510 + 410 + 650}{4} = \frac{2050}{4} = 512.5$$
$$\overline{wage}|(\text{public}_i = 1) = \frac{320 + 290 + 170 + 370}{4} = \frac{1150}{4} = 287.5$$

We observe that the mean wage for individuals in the public sector is 287.5, and for individuals in the private sector is 512.5. $\overline{wage}|(\text{public}_i = 0) = 512.5$, which is the same as the intercept $\hat{\beta}_0$. $\overline{wage}|(\text{public}_i = 1) = 287.5$, which is equal to the sum of the intercept and the slope: $\hat{\beta}_0 + \hat{\beta}_1 = 512.5 + (-225.0) = 287.5$. We can understand this better if we examine the estimated model in the two cases: when $\text{public}_i = 0$ and when $\text{public}_i = 1$. This gives us:

$$\widehat{wage}|(\text{public}_i = 0) = \hat{\beta}_0 + \hat{\beta}_1 \cdot 0 = \hat{\beta}_0 = 512.5$$
$$\widehat{wage}|(\text{public}_i = 1) = \hat{\beta}_0 + \hat{\beta}_1 \cdot 1 = \hat{\beta}_0 + \hat{\beta}_1 = 287.5$$

We can also estimate the model:

$$wage_i = \beta_0 + \beta_1\text{private}_i + u_i$$

which yields the following results:

Model 2: OLS, using observations 1–8
Dependent variable: wage

	Coefficient		Std. Error		*t*-ratio		*p*-value	
const	287	500	46	6146	6	168	0	0008***
private	225	000	65	9229	3	413	0	0143**

Mean dependent var	400.0000	S.D. dependent var	148.0347
Sum squared resid	52150.00	S.E. of regression	93.22911
R^2	0.660039	Adjusted R^2	0.603379
$F(1, 6)$	11.64909	P-value (F)	0.014265
Log-likelihood	−46.48126	Akaike criterion	96.96252
Schwarz criterion	97.12140	Hannan–Quinn	95.89092

If $\text{public}_i = 0$, we have $\hat{Y} = \hat{\beta}_0 = 278.5$, and if $\text{public}_i = 1$, we get $\hat{Y} = \hat{\beta}_0 + \hat{\beta}_1 = 278.5 + 225 = 512.5$. These are the same mean values as when we estimated the model $wage_i = \beta_0 + \beta_1 \text{public}_i + u_i$.

Additionally, we can estimate the model:

$$wage_i = \beta_1 \text{public}_i + \beta_2 \text{private}_i + u_i$$

which yields the following results:

Model 3: OLS, using observations 1–8
Dependent variable: wage

	Coefficient		Std. Error		*t*-ratio		*p*-value	
public	287	500	46	6146	6	168	0	0008***
private	512	500	46	6146	10	99	0	0000***

Mean dependent var	400.0000	S.D. dependent var	148.0347
Sum squared resid	52150.00	S.E. of regression	93.22911
R^2	0.660039	Adjusted R^2	0.603379
$F(1, 6)$	11.64909	P-value(F)	0.014265
Log-likelihood	−46.48126	Akaike criterion	96.96252
Schwarz criterion	97.12140	Hannan–Quinn	95.89092

where we see that if $\text{public}_i = 0$, we have $\hat{Y} = \hat{\beta}_1 = 278.5$, and if $\text{public}_i = 1$, we get $\hat{Y} = \hat{\beta}_2 = 512.5$, these are the same mean values as before. So, these are three ways to estimate the same effects.

If the dummy variable is equal to zero, the expected value of the dependent variable will be equal to the estimated value of the intercept, $\hat{\beta}_0$. However, if the dummy variable is equal to 1, we must add the slope coefficient $\hat{\beta}_1$. These estimates correspond to the mean values, since there are no other explanatory variables in the model. We can also use simple regression with the dummy variable private_i to find the same.

The interpretation of $\hat{\beta}_1$ for a dummy variable can be done as in simple regression with a regular continuous variable as in Chapter 5: If the explanatory variable increases by 1, the dependent variable changes by $\hat{\beta}_1$. The difference is that if a dummy variable increases by 1, this implies that we are comparing two categories (the units with the dummy variable equal to 1 and the units with the dummy variable equal to zero). Simple regression with a dummy variable can thus be used to calculate the mean in the two groups.

Interpreting $\hat{\beta}_1$ in Simple Regression with a Dummy Variable

If we have the model $Y_i = \beta_0 + \beta_1 D_i + u_i$ where D_i is a dummy variable, the interpretation of $\hat{\beta}_1$ is that it describes the change in Y_i when D_i increases by 1.

Additionally, we can interpret $\hat{\beta}_0$ as the expected value of the dependent variable when $D_i = 0$ and $\hat{\beta}_0 + \hat{\beta}_1$ as the expected value of the dependent variable when $D_i = 1$.

7.2 Generate Dummy Variables in Gretl and R

In addition to manually creating a dummy variable, we can construct a dummy variable based on a criterion. Often, we might have a variable country_i with values like "France", "Switzerland", "Germany", or the respective country codes 250, 756, and 276. We then often create a dummy variable for each country that takes on the values 1 or 0.

In Gretl, in the main window, you can click on the variable `country` and go to `Add` > `Dummies for discrete variable...` or right-click on the variable `country` and choose `Dummify...`. Then, you can select `Encode all values` to create a dummy variable for each possible value (in this case, each possible country code) or choose to exclude the observation with the lowest or highest number (since you often have to omit a variable to avoid perfect multicollinearity).

After creating dummy variables for each country in Gretl, you will have a set of dummy variables, as shown in Fig. 7.1, for the countries you have observations for (variables like `Dcountry_250`, `Dcountry_756`, and `Dcountry_276`), in addition to the original variable `country`.

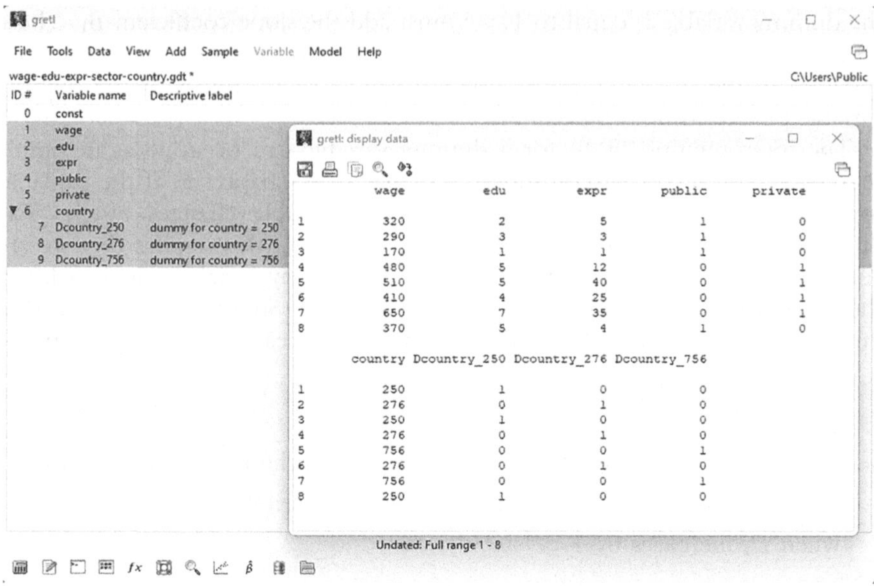

Fig. 7.1 Generate dummy variables in Gretl

In R, you can create a dummy variable based on a criterion using the `ifelse` function. To create a dummy variable for country code 250, "France", from the `country` variable, as shown in Fig. 7.1, you can write `France < - ifelse(country == 250, 1, 0)`. This code creates the variable `France`, which takes the value 1 if the `country` variable has the value 250, and 0 otherwise.

R code 7.1: Creating Dummy Variables in R

```
France <- ifelse(country == 250, 1, 0)
Switzerland <- ifelse(country == 756, 1, 0)
Germany <- ifelse(country == 276, 1, 0)
```

We then get what is shown in Table 7.2.

Table 7.2 Wage, education, experience, sector, and country

Wage	*Edu*	*Expr*	*Public*	*Private*	*Country*	*France*	*Switz*	*Ger*
320	2	5	1	0	250	1	0	0
290	3	3	1	0	276	0	0	1
170	1	1	1	0	250	1	0	0
480	5	12	0	1	276	0	0	1
510	5	40	0	1	756	0	1	0
410	4	25	0	1	276	0	0	1
650	7	35	0	1	756	0	1	0
370	5	4	1	0	250	1	0	0

7.3 Fixed Effects

A common way to use dummy variables is to add them to a simple or multiple regression model. For example, we can have a model like

$$Y_i = \beta_0 + \beta_1 X_i + \beta_2 D_i + u_i$$

where D_i is a dummy variable that can take the values $D_i = 0$ or $D_i = 1$, and Y_i and X_i are regular continuous or discrete variables. This leads to a multiple regression model with two explanatory variables: X_i and D_i. We estimate this model as

$$\hat{Y}_i = \hat{\beta}_0 + \hat{\beta}_1 X_i + \hat{\beta}_2 D_i$$

If $D_i = 0$, we get

$$\hat{Y}_i = \hat{\beta}_0 + \hat{\beta}_1 X_i$$

And if $D_i = 1$, we have

$$\begin{aligned} \hat{Y}_i &= \hat{\beta}_0 + \hat{\beta}_1 X_i + \hat{\beta}_2 \\ &= \underbrace{\hat{\beta}_0 + \hat{\beta}_2}_{\text{constant term}} + \hat{\beta}_1 X_i \end{aligned}$$

From this, we see that the constant term, when $D_i = 1$, becomes $\hat{\beta}_0 + \hat{\beta}_2$, whereas, as in simple regression, it is $\hat{\beta}_0$ when $D_i = 0$. So, we obtain two different estimated models depending on whether the dummy variable has a value of 0 or 1. This implies that we have a fixed effect that belongs to the group where the dummy variable has a value of 1 relative to the group where the dummy variable is equal to 0.

7.3.1 *Example: Wage, Education, and Sector*

By including the dummy variable public_i in the simple regression model with wage and education, $wage_i = \beta_0 + \beta_1 edu_i + u_i$, we obtain a multiple regression model with one dummy variable:

$$wage_i = \beta_0 + \beta_1 edu_i + \beta_2 \text{public}_i + u_i$$

We estimate this model in Gretl and obtain the following results:

Model 4: OLS, using observations 1–8
Dependent variable: wage

	Coefficient		Std. Error		*t*-ratio		*p*-value	
const	222	778	71	1241	3	132	0	0259**
edu	55	1852	12	7872	4	316	0	0076***
public	−87	0370	46	1048	−1	888	0	1177

Mean dependent var	400.0000	S.D. dependent var	148.0347
Sum squared resid	11037.04	S.E. of regression	46.98305
R^2	0.928051	Adjusted R^2	0.899271
$F(2, 5)$	32.24664	P-value(F)	0.001389
Log-likelihood	−40.26979	Akaike criterion	86.53958
Schwarz criterion	86.77790	Hannan–Quinn	84.93218

The estimated model is as follows:

$$\hat{wage}_i = \hat{\beta}_0 + \hat{\beta}_1 edu_i + \hat{\beta}_2 \cdot \text{public}_i$$
$$\hat{wage}_i = 222.8 + 55.2 edu_i - 87.0 \cdot \text{public}_i$$

This implies that $\hat{\beta}_0 = 222.8$, $\hat{\beta}_1 = 55.2$, and $\hat{\beta}_2 = 87.0$.

For employees in the private sector ($\text{public}_i = 0$), we have:

$$\hat{wage}_i = \hat{\beta}_0 + \hat{\beta}_1 edu_i$$
$$\hat{wage}_i = 222.8 + 55.2 edu_i$$

And for employees in the public sector ($\text{public}_i = 1$), we have:

$$\hat{wage}_i = \hat{\beta}_0 + \hat{\beta}_1 edu_i + \hat{\beta}_2$$
$$\hat{wage}_i = (\hat{\beta}_0 + \hat{\beta}_2) + \hat{\beta}_1 edu_i$$
$$\hat{wage}_i = (222.8 - 87.0) + 55.2 edu_i$$
$$\hat{wage}_i = 135.7 + 55.2 edu_i$$

From this, we can see that the intercept is 222.8 for the private sector and 135.7 for the public sector. The expected wage for employees with no education beyond high school ($edu_i = 0$) is lower for public sector employees than

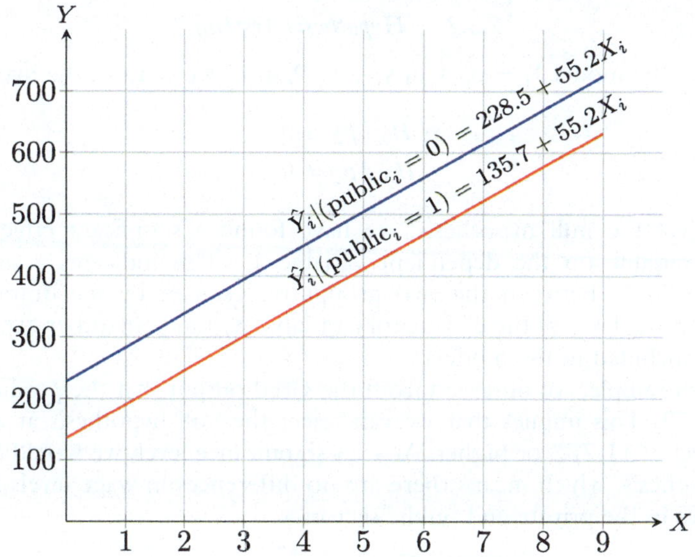

Fig. 7.2 Regression lines when using a fixed effect

for private sector employees. Additionally, the regression line will be higher for public sector employees than for private sector employees for all possible values of the explanatory variable, education. This can be illustrated by sketching the regression lines for $\text{public}_i = 0$ and $\text{public}_i = 1$, representing employees in the private and public sectors, respectively.

As shown in Fig. 7.2, the regression lines for the public sector (red) and private sector (blue) are parallel, but the line for the public sector is lower than the line for the private sector. The distance between the lines corresponds to the estimated coefficient for the dummy variable, $\hat{\beta}_2 = -87.0$.

In addition to the changed intercept indicating that the wage for an employee in the public sector with no education beyond high school is 87 lower than an employee in the private sector with the same educational level, there will also be a difference of 87 for all possible values of the explanatory variable education. Therefore, the wage will be 87 lower for public sector employees, assuming the number of years of education is the same.

Interpretation of $\hat{\beta}_1$ for a dummy variable

If we have the model $Y_i = \beta_0 + \beta_1 D_i + \beta_2 X_i + u_i$ where D_i is a dummy variable, the interpretation of $\hat{\beta}_1$ is that it describes the expected difference between the group where $D_i = 1$ and the group where $D_i = 0$, holding X_i (and any other explanatory variables) constant.

7.3.2 *Hypothesis Testing*

If we have the model $Y_i = \beta_0 + \beta_1 X_i + \beta_2 D_i + u_i$, we can test the hypotheses:

$$H_0 : \beta_2 = 0$$
$$H_A : \beta_2 \neq 0$$

If we reject the null hypothesis, we have found a significant effect of the dummy variable on the dependent variable Y_i. This indicates a significant difference in Y_i between the two groups represented by the dummy variable, given the level of the explanatory variable X_i (and any other explanatory variables included in the model).

In our example, we observed from the Gretl output that the *p*-value for $\hat{\beta}_2$ was 0.1177. This implies that we can reject the null hypothesis at a significance level of 11.77% or higher. At a 5% significance level, we would keep the null hypothesis, which means there are no differences in wage levels between employees in the private and public sectors.

7.4 Interaction Effects

Another common use of dummy variables is to employ them in an interaction term. This allows us to estimate interaction effects. We can achieve this by multiplying the dummy variable by an explanatory variable (in addition to including the explanatory variable), resulting in the estimated model:

$$Y_i = \beta_0 + \beta_1 X_i + \beta_2 D_i \cdot X_i + u_i$$

where the interaction term is $\beta_2 D_i \cdot X_i$.

If $D_i = 0$, we obtain the model:

$$Y_i = \beta_0 + \beta_1 X_i + u_i$$

And if $D_i = 1$, we get the model:

$$Y_i = \beta_0 + \beta_1 X_i + \beta_2 X_i + u_i$$
$$Y_i = \beta_0 + \underbrace{(\beta_1 + \beta_2)}_{\text{slope}} X_i + u_i$$

In other words, we have two simple regression models with different slopes. If $D_i = 0$, we have a slope of β_1, and if $D_i = 1$, we have a slope of $(\beta_1 + \beta_2)$. Therefore, the effect of the explanatory variable on the dependent variable differs for the two groups separated by the dummy variable.

7.4.1 *Example: Wage, Education, and Sector*

We are now using the dummy variable `public` to estimate the interaction effect between education and sector affiliation on wage. The model we wish to estimate is:

$$\text{wage}_i = \beta_0 + \beta_1 \text{edu}_i + \beta_2 \text{public}_i \cdot \text{edu}_i + u_i$$

To estimate this model, we need to treat the product of the explanatory variable education and the dummy variable for the public sector as a separate variable. In Gretl, we must generate a new variable $\text{edupublic}_i = \text{public}_i \cdot \text{edu}_i$. We can do this by going to `Add > Define new variable...` and entering `edupublic` = `public * edu`. This creates the new variable edupublic_i. The dataset used is shown in Table 7.3, where we see that the variable takes the value of edu_i when $\text{public}_i = 1$ and 0 otherwise.

By estimating the model with education and the new variable (the interaction term between education and the public sector dummy), we obtain:

Model 5: OLS, using observations 1–8
Dependent variable: wage

	Coefficient		Std. error		*t*-ratio		*p*-value	
const	160	069	37	6591	4	250	0	0081***
edu	67	6395	7	77151	8	704	0	0003***
edupublic	−22	2744	8	10999	−2	747	0	0405**

Mean dependent var	400.0000	S.D. dependent var	148.0347
Sum squared resid	7535.313	S.E. of regression	38.82090
R^2	0.950878	Adjusted R^2	0.931229
$F(2, 5)$	48.39371	P-value(F)	0.000535
Log-likelihood	−38.74316	Akaike criterion	83.48633
Schwarz criterion	83.72465	Hannan–Quinn	81.87893

Table 7.3 Dataset with generated interaction term

i	*Wage*	*Edu*	*Public*	*Edupublic*
1	320	2	1	2
2	290	3	1	3
3	170	1	1	1
4	480	5	0	0
5	510	5	0	0
6	410	4	0	0
7	650	7	0	0
8	370	5	1	5

The estimated model is then:

$$\widehat{wage}_i = \hat{\beta}_0 + \hat{\beta}_1\text{edu}_i + \hat{\beta}_2\text{edu}_i \cdot \text{public}_i$$
$$\widehat{wage}_i = 160.1 + 67.6\text{edu}_i - 22.3\text{edu}_i \cdot \text{public}_i$$

where $\hat{\beta}_0 = 160.1$, $\hat{\beta}_1 = 67.6$, and $\hat{\beta}_2 = -22.3$.

For employees in the private sector ($\text{public}_i = 0$), we have:

$$\widehat{wage}_i = \hat{\beta}_0 + \hat{\beta}_1\text{edu}_i$$
$$\widehat{wage}_i = 160.1 + 67.6\text{edu}_i$$

And for employees in the public sector ($\text{public}_i = 1$), we get:

$$\widehat{wage}_i = \hat{\beta}_0 + \hat{\beta}_1\text{edu}_i + \hat{\beta}_2\text{edu}_i$$
$$\widehat{wage}_i = \hat{\beta}_0 + (\hat{\beta}_1 + \hat{\beta}_2)\text{edu}_i$$
$$\widehat{wage}_i = 160.1 + (67.6 - 22.3)\text{edu}_i$$
$$\widehat{wage}_i = 160.1 + 45.4\text{edu}_i$$

The regression lines for employees in the private and the public sectors are shown in (Fig. 7.3).

Interpretation of the Coefficient for the Interaction Term

In the model $Y_i = \beta_0 + \beta_1 X_i + \beta_2 D_i \cdot X_i + u_i$, where D_i is a dummy variable, $D_i \cdot X_i$ represents an interaction term. The interpretation of $\hat{\beta}_2$ describes the difference in the effect of X_i on Y_i (i.e., the slope) between the two groups where $D_i = 1$ and where $D_i = 0$.

7.4.2 Hypothesis Testing

In the model $Y_i = \beta_0 + \beta_1 X_i + \beta_2 D_i \cdot X_i + u_i$, we can use the hypotheses

$$H_0 : \beta_2 = 0$$
$$H_A : \beta_2 \neq 0$$

to test whether there is a difference between the groups ($D_i = 0$ and $D_i = 1$) in how the explanatory variable affects the dependent variable.

In our example, $\text{wage}_i = \beta_0 + \beta_1\text{edu}_i + \beta_2\text{public}_i \cdot \text{edu}_i + u_i$, we found a *p*-value of 0.0405 for $\hat{\beta}_2$. This means that we reject the null hypothesis at a 4.05% significance level or lower, which implies that we have found a significant difference in how education affects wages for the public and private sectors.

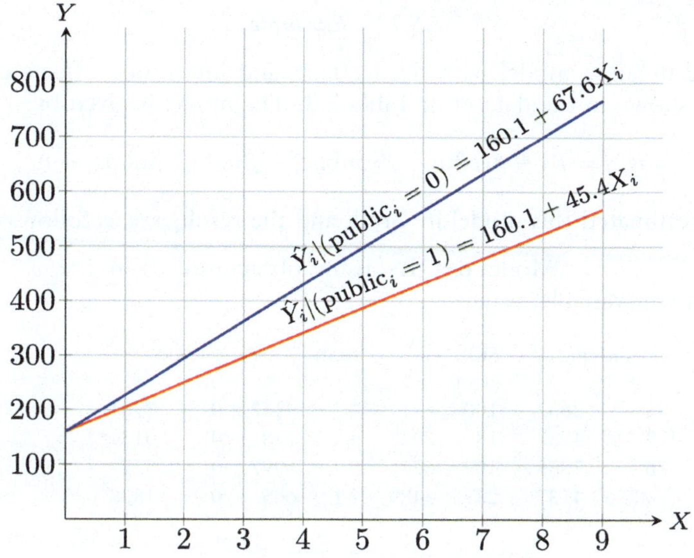

Fig. 7.3 Regression lines when using an interaction term

7.5 Both Fixed Effects and Interaction Terms

By including both the fixed effect by using the dummy variable separately and including the interaction term, we can estimate both the fixed effect and the interaction effect. We have the model:

$$Y_i = \beta_0 + \beta_1 X_i + \beta_2 D_i + \beta_3 X_i \cdot D_i + u_i$$

The estimated model is:

$$\hat{Y}_i = \hat{\beta}_0 + \hat{\beta}_1 X_i + \hat{\beta}_2 D_i + \hat{\beta}_3 X_i \cdot D_i$$

If $D_i = 0$, we have:

$$\hat{Y}_i = \hat{\beta}_0 + \hat{\beta}_1 X_i$$

And if $D_i = 1$, we have:

$$\hat{Y}_i = \hat{\beta}_0 + \hat{\beta}_1 X_i + \hat{\beta}_2 + \hat{\beta}_3 X_i$$
$$\hat{Y}_i = \underbrace{(\hat{\beta}_0 + \hat{\beta}_2)}_{\text{constant term}} + \underbrace{(\hat{\beta}_1 + \hat{\beta}_3)}_{\text{slope}} X_i$$

There are different constant terms and different slopes for the two groups. This means that we are estimating both fixed effects and interaction effects simultaneously.

7.5.1 *Example*

We can estimate a model with fixed effects and interaction effects using the variables shown in the dataset in Table 7.3. The model is given by:

$$\text{wage}_i = \beta_0 + \beta_1 \text{edu}_i + \beta_2 \text{public}_i + \beta_3 \text{edu}_i \cdot \text{public}_i + u_i$$

We have estimated this model in Gretl, and the results are as follows:

Model 6: OLS, using observations 1–8
Dependent variable: wage

	Coefficient		Std. error		*t*-ratio		*p*-value	
const	95	2632	100	791	0	9452	0	3981
edu	79	4737	18	7977	4	228	0	0134**
public	76	7368	109	677	0	6997	0	5227
edupublic	−37	4737	23	3489	−1	605	0	1838

Mean dependent var	400.0000	S.D. dependent var	148.0347
Sum squared resid	6713.684	S.E. of regression	40.96854
R^2	0.956234	Adjusted R^2	0.923410
$F(3, 4)$	29.13181	P-value(F)	0.003539
Log-likelihood	−38.28135	Akaike criterion	84.56271
Schwarz criterion	84.88048	Hannan–Quinn	82.41950

The estimated model is as follows:

$$\widehat{wage}_i = \hat{\beta}_0 + \hat{\beta}_1 \text{edu}_i + \hat{\beta}_2 \text{public}_i + \hat{\beta}_3 \text{edu}_i \cdot \text{public}_i$$
$$\widehat{wage}_i = 95.3 + 79.5 \text{edu}_i + 76.7 \text{public}_i - 37.5 \text{edu}_i \cdot \text{public}_i$$

If $\text{public}_i = 0$, we get:

$$\widehat{wage}_i = \hat{\beta}_0 + \hat{\beta}_1 \text{edu}_i$$
$$\widehat{wage}_i = 95.3 + 79.5 \text{edu}_i$$

And if $\text{public}_i = 1$, we have:

$$\widehat{wage}_i = \hat{\beta}_0 + \hat{\beta}_1 \text{edu}_i + \hat{\beta}_2 + \hat{\beta}_3 \text{edu}_i$$
$$\widehat{wage}_i = \underbrace{(\hat{\beta}_0 + \hat{\beta}_2)}_{\text{intercept}} + \underbrace{(\hat{\beta}_1 + \hat{\beta}_3)}_{\text{slope}} \text{edu}_i$$
$$\widehat{wage}_i = (95.3 + 76.7) + (79.5 - 37.5) \text{edu}_i$$
$$\widehat{wage}_i = 172.0 + 42.0 \text{edu}_i$$

The two estimated relationships are graphically shown in Fig. 7.4.

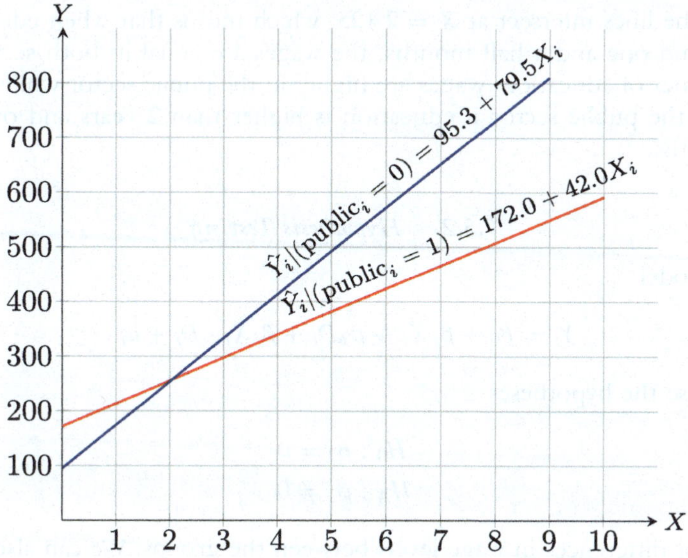

Fig. 7.4 Regression lines using both fixed and interaction effects

From Fig. 7.4, we can see that for individuals with very low education, those employed in the public sector have a higher wage than those in the private sector (corresponding to intercepts of 95.3 and 172.0, respectively, if $\text{edu}_i = 0$). However, since the line for the private sector is steeper (due to a higher slope) than that for the public sector (slopes of 79.5 and 42.0, respectively), wages increase more for each additional year of education for individuals employed in the public sector than for those in the private sector. We observe that the lines intersect at approximately $X = 2$, meaning that the estimated wage for individuals with 2 years of education will be roughly the same in both the public and private sectors:

$$\widehat{wage}_i|(\text{public}_i = 0, edu = 2) = 95.3 + 79.5 \cdot 2 = 254.3$$
$$\widehat{wage}_i|(\text{public}_i = 1, edu = 2) = 172.0 + 42.0 \cdot 2 = 256.0$$

We can also find the exact point of intersection by determining the value of X (education) at which the two lines are equal:

$$\widehat{wage}_i|(\text{public}_i = 0) = \widehat{wage}_i|(\text{public}_i = 1)$$
$$95.3 + 79.5X = 172.0 + 42.0X$$
$$79.5X - 42X = 172.0 - 95.3$$
$$37.5X = 76.7$$
$$X = 2.125$$

Hence, the lines intersect at $X = 2.125$, which means that when education is 2 years and one and a half months, the wages are equal in both sectors. For lower values of education, wages are higher in the public sector while they are lower in the public sector if education is higher than 2 years and one and a half months.

7.5.2 *Hypothesis Testing*

In the model

$$Y_i = \beta_0 + \beta_1 X_i + \beta_2 D_i + \beta_3 X_i \cdot D_i + u_i$$

we can use the hypotheses:

$$H_0 : \beta_2 = 0$$
$$H_A : \beta_2 \neq 0$$

to test for differences in wage levels between the groups. We can also use the hypotheses:

$$H_0 : \beta_3 = 0$$
$$H_A : \beta_3 \neq 0$$

to test for differences in how the explanatory variable affects the dependent variable between the groups.

From the Gretl output, we see that the *p*-values for the estimated coefficients of `public` and `edu` · `public` are high (0.52 and 0.18, respectively). This means that we do not find fixed effects between the groups or interaction effects.

In addition, we can use an *F*-test to test the hypotheses:

$$H_0 : \beta_2 = 0 \text{ and } \beta_3 = 0$$
$$H_A : \beta_2 \neq 0 \text{ and/or } \beta_3 \neq 0$$

This allows us to test whether there is a difference between the groups without considering whether the difference lies in the level of the dependent variable or in how the explanatory variable affects the dependent variable.

To do this in Gretl, go to `Test` > `Omit variables` and select `public` and `edu` · `public` as the variables to omit. This will give you:

```
Test on Model 6:

Null hypothesis:
the regression parameters are zero for the variables
public, edupublic
Test statistic: F(2, 4) = 3.63144, p-value 0.126131
Omitting variables improved 0 of 3 information criteria.
```

The p-value of 0.12 tells us that we cannot reject the null hypothesis at a significance level of 12% or lower. Therefore, we have not found a significant difference between the groups at conventional significance levels.

In R, you can perform this test by estimating the model with restrictions and then conducting the test using the `anova` function, as shown in R code 6.2. This will yield the same outcome as the test in Gretl.

7.5.3 Example with Wage and City

To estimate how wage is influenced by the country in which an employee lives, we can use the model:

$$\text{wage}_i = \beta_0 + \beta_1 \text{Switzerland}_i + \beta_2 \text{Germany}_i + u_i \tag{7.1}$$

In this model, we have a wage model where the dummy variable for Switzerland (Switzerland_i) and the dummy variable for Germany (Germany_i) are explanatory variables. This means that France is the reference category. We could also include other explanatory variables to control for factors other than the place of residence, such as education, experience, and sector.

By estimating this model, we obtain:

Model 7: OLS, using observations 1–8
Dependent variable: wage

	Coefficient		Std. Error		*t*-ratio		*p*-value	
const	286	667	57	6965	4	969	0	0042***
Germany	106	667	81	5952	1	307	0	2480
Switzerland	293	333	91	2262	3	215	0	0236**

Mean dependent var	400.0000	S.D. dependent var	148.0347
Sum squared resid	49933.33	S.E. of regression	99.93331
R^2	0.674489	Adjusted R^2	0.544285
$F(2, 5)$	5.180240	P-value(F)	0.060452
Log-likelihood	−46.30752	Akaike criterion	98.61504
Schwarz criterion	98.85336	Hannan–Quinn	97.00763

We have thus estimated the intercept to be $\hat{\beta}_0 = 286.67$, the slope for the dummy variable for Switzerland as $\hat{\beta}_1 = 293.33$, and the slope for Germany as $\hat{\beta}_2 = 106.67$. We can observe from the p-values that Switzerland has a significant effect at a 5% significance level, but there is no significant effect in Germany compared to France. Additionally, we can test whether there is any difference in mean wage between the countries through a multiple hypothesis test:

$$H_0 : \beta_1 = \beta_2 = 0$$
$$H_A : \beta_1 \neq 0 \text{ and/or } \beta_2 \neq 0$$

Under the null hypothesis, the dummy variables for Switzerland and Germany do not have a significant effect, which implies that the mean wages are the same in all three countries. However, under the alternative hypothesis, at least one of the countries is different.

Manual F-Test

To perform the hypothesis test described above using an F-test, we start with the model (7.1). Under the null hypothesis, we set $\beta_1 = \beta_2 = 0$, resulting in the model:

$$\text{wage}_i = \beta_0 + u_i$$

This model has only an intercept, which leads to an $R^2_{\text{null}} = 0$. Therefore, we can calculate the test statistic as follows:

$$F = \frac{(R^2 - R^2_{\text{null}})/m}{(1 - R^2)/(n - k)} = \frac{(0.674 - 0)/2}{(1 - 0.674)/(8 - 3)} = \frac{0.337}{0.0652} = 5.169$$

To find the critical value, we consult the F-distribution with $m = 2$ degrees of freedom in the numerator and $n - k = 8 - 3 = 5$ degrees of freedom in the denominator. This critical value is 5.786. Since the test statistic is less than the critical value, we keep the null hypothesis. We cannot accept the claim that wages in the countries are different.

F-*Test in Gretl*

By navigating to `Tests >  Omit variables` after estimating (7.1), we obtain the following output:

```
Test for omission of variables -
Null hypothesis: parameters are zero for the variables
Germany
Switzerland
Test statistic: F(2, 5) = 5.18024
with p-value = P(F(2, 5) > 5.18024) = 0.0604523
```

The test statistic is indeed 5.18, which is close to what we manually calculated, with a p-value of 0.060. This implies that we keep the null hypothesis at a 5% significance level.

F-Test in R

To perform the F-test in R, we use the anova function, as shown in Sect. 6.4. First, we estimate the model under the null hypothesis, $Y_i = \beta_0 + u_i$, which is the model with only a constant term. Then, we estimate the model with the dummy variables for France and Switzerland.

R code 7.2: F-test in R

```
H0model <- lm(wage ~ 1)
SwitzerlandGermany <- lm(wage ~ switzerland + germany)
anova(SwitzerlandGermany, H0model)
```

The output will look like this:

```
Analysis of Variance Table

Model 1: wage ~ switzerland + germany
Model 2: wage ~ 1
Res.Df    RSS Df Sum of Sq      F  Pr(>F)
1      5  49933
2      7 153400 -2   -103467 5.1802 0.06045 .
---
Signif. codes:
0 '***' 0.001 '**' 0.01 '*' 0.05 '.' 0.1 ' ' 1
```

The test statistic is $F = 5.18$, similar to what we manually calculated, with a p-value of 0.060. This indicates that we keep the null hypothesis at a 5% significance level.

7.6 Dummy Variable as Dependent Variable

When dealing with a regression model where the dependent variable is a dummy variable, using linear regression can be problematic. For instance, in the model

$$D_i = \beta_0 + \beta_1 X_i + u_i$$

$\hat{\beta}_1$ would represent the expected change in D_i when X_i increases by 1. Since D_i can only take values 0 or 1 (or other binary values), it does not make sense to talk about D_i increasing by a specific amount. Typically, we are more interested in the probability of D_i being 1 or 0.

A more appropriate method for modeling binary dependent variables is logistic regression. Logistic regression estimates the probability of D_i being 1 as a function of X_i, and it can provide insights into how changes in X_i affect the probability of D_i being 1 or 0.

7.6.1 Linear Probability Models

We can interpret $\hat{\beta}_1$ as the expected effect on D_i when X_i increases by 1. However, we have a linear probability model that might have limited explanatory power. Since the dependent variable can only take two possible values, there is often a low correlation between the dummy variable D_i and the continuous variable X_i.

In the model

$$\text{private}_i = \beta_0 + \beta_1 \text{edu}_i + u_i$$

we can estimate how the level of education affects the probability that a given individual works in the private sector relative to the public sector. Still, this model often has limited explanatory power, and it is challenging to interpret the estimated slope. This limitation can be observed in Fig. 7.5, where there are significant deviations between the regression line $\widehat{\text{private}}_i = \hat{\beta}_0 + \hat{\beta}_1 \text{edu}_i = -0.269 + 0.192 \cdot \text{edu}_i$ and the observations. The regression output shown below indicates that the model has an explanatory power of only about 48%.

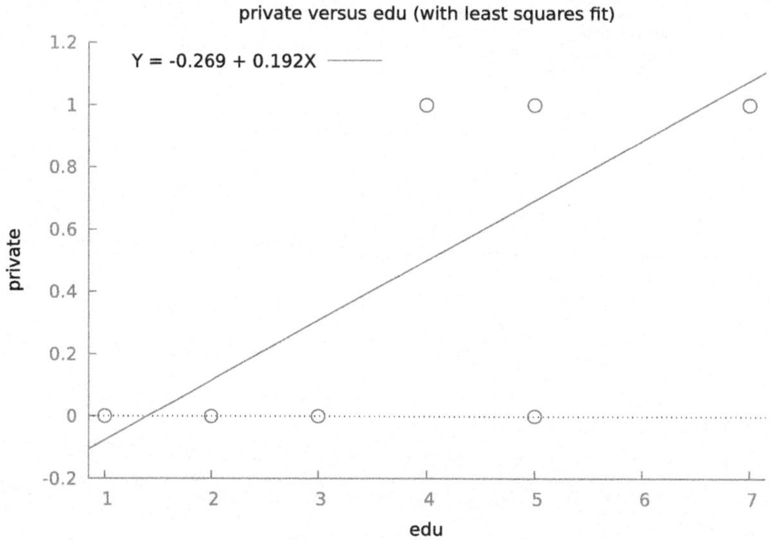

Fig. 7.5 Linear Probability Model

Model 8: OLS, using observations 1–8
Dependent variable: private

	Coefficient		Std. error		*t*-ratio		*p*-value	
const	−0	269231	0	357971	−0	7521	0	4804
edu	0	192308	0	0815892	2	357	0	0565*

Mean dependent var	0.500000	S.D. dependent var	0.534522
Sum squared resid	1.038462	S.E. of regression	0.416025
R^2	0.480769	Adjusted R^2	0.394231
$F(1, 6)$	5.555556	P-value(F)	0.056513
Log-likelihood	−3.184703	Akaike criterion	10.36941
Schwarz criterion	10.52829	Hannan–Quinn	9.297804

If a person has 5 years of education, the model will estimate the probability of that person being employed in the public sector as:

$$Pr(Y|X = 5) = \hat{\beta}_0 + \hat{\beta}_1 \cdot 5 = -0.269 + 0.192 \cdot 5 = 0.591$$

7.6.2 *Probit Models*

A linear probability model can be represented as

$$Y_i = \beta_0 + \beta_1 X_{1i} + \beta_2 X_{2i} + \ldots + \beta_k X_{ki} + u_i$$

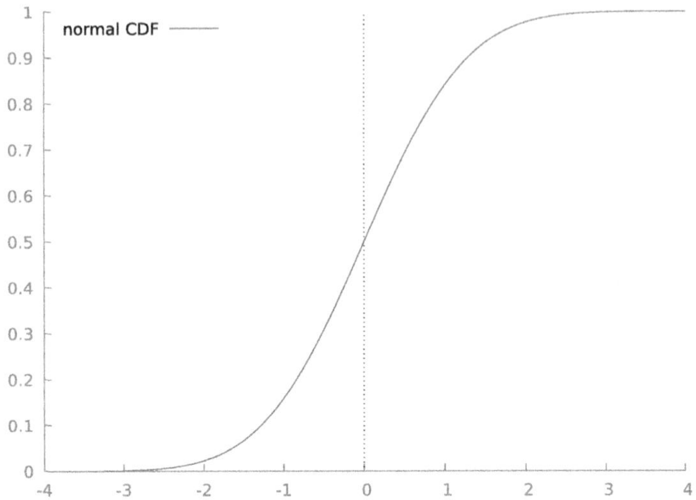

Fig. 7.6 Cumulative standard normal distribution

where Y_i is a binary variable. This can be expressed as:

$$Pr(Y = 1|X_1, X_2, \ldots, X_k) = \beta_0 + \beta_1 X_1 + \beta_2 X_2 + \ldots + \beta_k X_k$$

A probit model is written as:

$$Pr(Y = 1|X_1, X_2, \ldots, X_k) = \Phi(\beta_0 + \beta_1 X + \beta_2 X_2 + \ldots + \beta_k X_k) \quad (7.2)$$

Here, Φ represents the cumulative standard normal distribution. By inserting the estimated coefficients ($\hat{\beta}_0, \hat{\beta}_1, \ldots, \hat{\beta}_k$) and the value of X into the equation, we can determine the probability of $Y = 1$ and find the tail probability in the cumulative normal distribution. The cumulative standard normal distribution is illustrated in Fig. 7.6.

A simple probit model can be expressed as:

$$Pr(Y = 1|X) = \Phi(\beta_0 + \beta_1 X)$$

In our example, Y is the dummy variable `private`, which is equal to 1 if the person is employed in the private sector and 0 otherwise, and X represents education.

Probit Model in Gretl

To specify a probit model in Gretl, go to `Model > Limited dependent variable > Probit > Binary`. This will open a window where you can define the dependent variable and explanatory variables for a probit model. You can select `private` as the dependent variable (`Dependent variable`) and a constant term (`const`), as well as `edu` as the explanatory variable (`Regressor`).

You will then obtain:

Model 9: Probit, using observations 1–8
Dependent variable: private
Standard errors based on Hessian

	Coefficient		Std. Error		z		Slope*	
const	−3	58229	2	63450	−1	360		
edu	0	863573	0	588279	1	468	0	341705

Mean dependent var	0.500000	S.D. dependent var	0.534522
McFadden R^2	0.457378	Adjusted R^2	0.096704
Log-likelihood	−3.008936	Akaike criterion	10.01787
Schwarz criterion	10.17675	Hannan–Quinn	8.946269

*Evaluated at the mean
Number of cases "correctly predicted" = 6 (75.0%)
Likelihood ratio test: $\chi^2(1) = 5.072$ [0.0243]
Test for normality of residual–

Null hypothesis: error is normally distributed
Test statistic: $\chi^2(2) = 4.39171$
with p-value = 0.111263

As shown in Fig. 7.7, the estimates align along the cumulative standard normal distribution. This results in the model having less deviation between estimated and observed values for low and high values of the explanatory variable (education). Consequently, the probit model has better explanatory power compared to the linear probability model.

For an education level of 5 years ($X = 5$), the probability of a person being employed in the private sector ($Pr(Y = 1)$) is as follows:

$$Pr(Y = 1|X = 5) = \Phi(\hat{\beta}_0 + \hat{\beta}_1 X) = \Phi(-3.58 + 0.86 \cdot 5) = \Phi(0.72) = 0.764$$

To calculate $\Phi(0.72) = 0.764$, you can use the command `pr = cdf(N, 0.72)` in the Gretl console. This will yield the result `Generated scalar pr = 0.764238`. Therefore, there is a 76.4% probability that a person with 5 years of education is employed in the private sector.

To find the estimated probabilities for all observed values of the explanatory variable, you can go to `Analysis > Display actual, fitted, residual`. This will provide the probability of $Y = 1$ for all observed values of X.

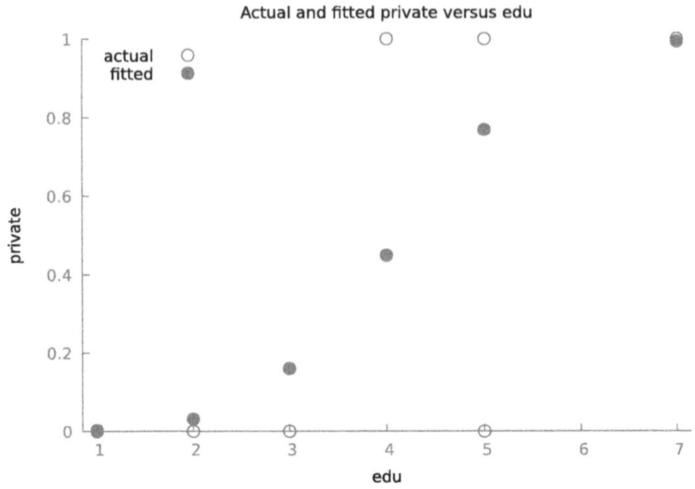

Fig. 7.7 Observations and predictions from the estimated probit model

Model estimation range: 1–8

	private	fitted	residual
1	0.000000	0.031788	−0.031788
2	0.000000	0.160704	−0.160704
3	0.000000	0.003277	−0.003277
4	1.000000	0.769006	0.230994
5	1.000000	0.769006	0.230994
6	1.000000	0.449076	0.550924
7	1.000000	0.993106	0.006894
8	0.000000	0.769006	−0.769006

Forecast evaluation statistics (T = 8)

Mean Error	0.0068789
Mean Squared Error	0.35856
Root Mean Squared Error	0.5988
Mean Absolute Error	0.24807
Theil's U_1	0.27197

Here we can see that for individuals with 5 years of education, i.e., $i = 4$, $i = 5$, and $i = 8$, there is a 76.9% probability of being employed in the private sector.

Probit Model in R

We can estimate a probit model in R by using the `glm` function. `glm` (Generalized Linear Model) is a function that contains several non-linear models, including the probit model.

R code 7.3: Probit model in R

General exapmle and our worked example

```
probitmodel <- glm(Y ~ X1 X2 X3,
family=binomial(link="probit"))
summary(probitmodel)

probitmodel <- glm(private ~ edu,
family=binomial(link="probit"))
summary(probitmodel)
```

By estimating the probit model in our example, we then get

```
Call:
glm(formula = private ~ edu,
family = binomial(link = "probit"))

Deviance Residuals:
Min          1Q    Median        3Q        Max
-1.7119  -0.3386   0.0183   0.7248   1.2653

Coefficients:
Estimate Std. Error z value Pr(>|z|)
(Intercept)  -3.5822       2.4864  -1.441     0.150
edu           0.8636       0.5619   1.537     0.124

(Dispersion parameter for binomial family taken to be 1)

Null deviance: 11.0904  on 7  degrees of freedom
Residual deviance:  6.0179  on 6  degrees of freedom
AIC: 10.018

Number of Fisher Scoring iterations: 7
```

By typing probitmodel$fitted.values, we obtain all the predicted probabilities for our observed values of the explanatory variable.

```
> probitmodel$fitted.values
1          2          3          4
0.03178973 0.16070783 0.00327713 0.76900564
5          6          7          8
0.76900564 0.44907906 0.99310534 0.76900564
```

From this, we can see that for individuals with 5 years of education, i.e., $i = 4$, $i = 5$, and $i = 8$, there is a 76.9

7.6.3 *Logit Models*

Another non-linear probability model which is widely used is the **logit model**. Instead of using the standard normal distribution like the probit model, the logit model employs the cumulative standard logistic probability distribution. The model is defined as

$$Pr(Y = 1|X_1, X_2, \ldots, X_k) = F(\beta_0 + \beta_1 X + \beta_2 X_2 + \cdots + \beta_k X_k) \qquad (7.3)$$

The only difference from the probit model is thus that we use F instead of Φ since it's based on a different probability distribution. We can also express this as

$$Pr(Y = 1|X_1, X_2, \ldots, X_k) = \frac{1}{1 + e^{-(\beta_0 + \beta_1 X_1 + \beta_2 X_2 + \cdots + \beta_k X_k)}}$$

This allows us to calculate the probability directly after estimating the coefficients.

Logit Models in Gretl
A simple logit model can be expressed as follows:

$$Pr(Y = 1|X) = F(\beta_0 + \beta_1 X) = \frac{1}{1 + e^{-(\beta_0 + \beta_1 X)}}$$

In our example, the dependent variable Y is the dummy variable `private`, which takes the value 1 if a person is employed in the private sector and 0 otherwise, and X is the level of education.

To estimate a logit model in Gretl, you can go to `Model > Limited dependent variable > Logit > Binary`. This will open a window where you can specify the dependent variable (`Dependent variable`) and the explanatory variables (`Regressor`). In our case, you can select `private` as the dependent variable, `const` as the constant term, and `edu` as the regressor.

We then get

Model 12: Logit, using observations 1–8
Dependent variable: private
Standard errors based on Hessian

	Coefficient		Std. error		z		Slope*	
const	−5	92361	4	49416	−1	318		
edu	1	43680	1	01500	1	416	0	356418

Mean dependent var	0.500000	S.D. dependent var	0.534522
McFadden R^2	0.450053	Adjusted R^2	0.089380
Log-likelihood	−3.049551	Akaike criterion	10.09910
Schwarz criterion	10.25799	Hannan–Quinn	9.027500

*Evaluated at the mean
Number of cases "correctly predicted" = 6 (75.0%)
Likelihood ratio test: $\chi^2(1) = 4.991$ [0.0255]

For an education level of 5 years ($X = 5$), the probability that a person is employed in the private sector ($Pr(Y = 1)$) is calculated as follows:

$$Pr(Y = 1|X = 5) = F(\hat{\beta}_0 + \hat{\beta}_1 X) = F(-5.92 + 1.44 \cdot 5) = F(1.28)$$
$$= \frac{1}{1 + e^{-1.28}} = \frac{1}{1 + 0.278} = 0.782$$

Therefore, according to the simple logit model we estimated, there is a 78.2% probability that a person with 5 years of education is employed in the private sector.

You can also obtain the estimated probabilities for all values of the explanatory variable by going to `Analysis > Display actual, fitted, residual`. This will provide the probabilities of $Y = 1$ for all observed values of X.

Model estimation range: 1–8

	private	fitted	residual
1	0.000000	0.045217	−0.045217
2	0.000000	0.166142	−0.166142
3	0.000000	0.011131	−0.011131
4	1.000000	0.779090	0.220910
5	1.000000	0.779090	0.220910
6	1.000000	0.456008	0.543992
7	1.000000	0.984233	0.015767
8	0.000000	0.779090	−0.779090

Forecast evaluation statistics (T = 8)

Mean Error	−7.1332e-015
Mean Squared Error	0.35891
Root Mean Squared Error	0.59909
Mean Absolute Error	0.25039
Theil's U_1	0.27144

At the bottom of the Gretl output, we see that a "Likelihood ratio test" has been conducted. This is a multiple hypothesis test with the null hypothesis that all explanatory variables in the logit model are equal to zero:

$$H_0 : \beta_1 = \beta_2 = \cdots = \beta_k = 0$$
$$H_A : \beta_1 \neq 0 \text{ and/or } \beta_2 \neq 0 \ldots \text{ and/or } \beta_k \neq 0$$

This test follows a chi-squared distribution, and we need to find the critical value from the chi-squared distribution with $k - 1$ degrees of freedom. In Gretl, you can find the critical value from the chi-squared distribution by going to "Tools > Statistical tables" and selecting the "chi-square" tab. Here, you can enter the degrees of freedom, `df`, and the right-tail probability, `right-tail probability`. In R, you can find the critical value using the formula `qchisq(alpha,df, lower.tail=FALSE`, as shown in R code 3.3.

At a significance level of 5%, we have $\chi^2_{0.05}(1) = 3.84146$. The test statistic of 4.991 is higher than this critical value, so we reject the null hypothesis.

Logit Model i R

In R, we can estimate a logit model using the `glm` function, just like we did for the probit model.

R code 7.4: Logit model in R

General example and our worked example

```
logitmodel <- glm(Y ~ X1 X2 X3,
family=binomial(link="logit"))
summary(logitmodel)

logitmodel <- glm(private ~ edu,
family=binomial(link="logit"))
summary(logitmodel)
```

By estimating the logit model in our example, we get

```
Call:
glm(formula = private ~ edu,
family = binomial(link = "logit"))

Deviance Residuals:
Min        1Q      Median    3Q        Max
-1.73781  -0.37886  0.01433  0.70658  1.25319

Coefficients:
              Estimate Std. Error z value Pr(>|z|)
(Intercept)   -5.924      4.494  -1.318    0.187
edu            1.437      1.015   1.416    0.157

(Dispersion parameter for binomial family taken to be 1)

Null deviance: 11.0904  on 7  degrees of freedom
Residual deviance:  6.0991  on 6  degrees of freedom
AIC: 10.099

Number of Fisher Scoring iterations: 6
```

By typing `logitmodel$fitted.values`, we can obtain the predicted probabilities for our observed values of the explanatory variable.

```
> logitmodel$fitted.values
1          2          3          4
0.04521677 0.16614191 0.01113118 0.77908974

5          6          7          8
0.77908974 0.45600765 0.98423326 0.77908974
```

From this, we can observe that for individuals with $i = 4$, $i = 5$, and $i = 8$, all of whom have 5 years of education, the expected probability of being employed in the private sector is 77.9%.

7.6.4 *Which Probability Model to Use?*

We have now examined three different models where we have a dummy variable as the dependent variable. That is, we wanted to estimate how one or more variables affect the probability of the dependent variable being equal to 1. There are advantages and disadvantages to the three models, both in terms of estimation methods and the interpretation of results.

While it is easy to interpret estimates in a linear regression model, there is a risk that the errors are not normally distributed, and we may have heteroscedasticity, as suggested by the large deviations in Fig. 7.5.

Both the probit and logit models take this into account, although it can be more challenging to interpret the estimated coefficients, especially in the probit model. Since probit and logit models use different probability distributions, the choice of the best model may depend on our observations. This can be investigated by comparing, for example, the computed log-likelihood for each model. The choice of method can also be determined by theoretical justifications.

CHAPTER 8

Non-Linear Relationships

Even though the ordinary least squares method is a linear model, we often have non-linear relationships between the variables. This is covered in this chapter through various forms of non-linearity using the ordinary least squares method. The focus is on logarithmic and semi-logarithmic models, as well as polynomial and inverse relationships.

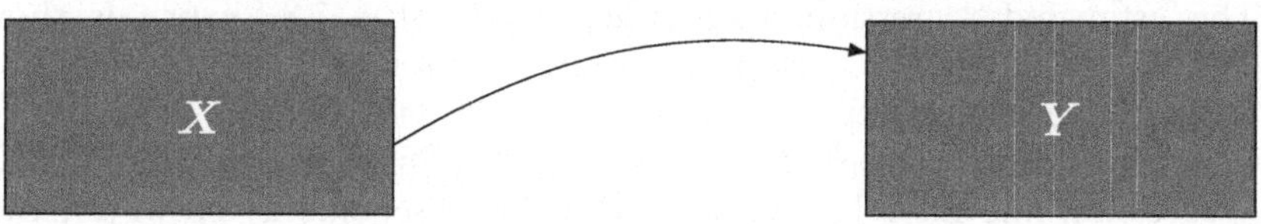

If we start with the dataset of eight individuals (as shown in Table 8.1) and aim to estimate a simple regression model for wage where experience is the only explanatory variable, we have the model:

$$\text{wage}_i = \beta_0 + \beta_1 \text{exper}_i + u_i \tag{8.1}$$

Supplementary Information The online version contains supplementary material available at https://doi.org/10.1007/978-3-031-53142-2_8.

B. Karlsen Kivedal, *Applied Statistics and Econometrics*, https://doi.org/10.1007/978-3-031-53142-2_8

Table 8.1 Wage and experience

i	*1*	*2*	*3*	*4*	*5*	*6*	*7*	*8*
wage	320	290	170	480	510	410	650	370
expr	5	3	1	12	40	25	35	4

These yield the following estimates:

Model 1: OLS, using observations 1–8
Dependent variable: wage

	Coefficient	Std. Error	*t*-ratio	*p*-value
const	277.355	46.9739	5.904	0.0010***
expr	7.84928	2.20066	3.567	0.0118**

Mean dependent var	400.0000	S.D. dependent var	148.0347
Sum squared resid	49,161.57	S.E. of regression	90.51848
R^2	0.679520	Adjusted R^2	0.626107
$F(1, 6)$	12.72194	p-value(F)	0.011830
Log-likelihood	−46.24521	Akaike criterion	96.49042
Schwarz criterion	96.64931	Hannan–Quinn	95.41882

The estimated regression line, $\widehat{\text{wage}}_i = 277.4 + 7.8 \cdot \text{expr}_i$, is shown in Fig. 8.1. The estimated slope indicates that for each additional year of experience, the wage is expected to increase by 7.849.

Experience explains nearly 68% of the variation in wage ($R^2 = 0.6795$), and the regression line fits relatively well with the observations. However, there is a relatively large deviation for the observation with the lowest experience (individual $i = 3$ with expr $= 1$ and wage $= 170$). The deviations also seem to be larger for individuals with extensive experience. This can be studied more closely by examining the residuals as shown in Fig. 8.2 and Table 8.2. Here we see that there are some significant deviations.

An alternative model for explaining wage with experience is the **non-linear model** $\text{wage}_i = \beta_0 + \beta_1 \cdot \ln(expr_i)$. The estimated model is shown in Figure 8.3. This model may fit the observations better than the linear model and thus may be better suited to estimate the effect of experience on wage.

The non-linear model in Fig. 8.3 is a lin-log model which will be presented in the next subsection. Furthermore, several other types of non-linear models suitable for estimating non-linear relationships between variables will also be presented.

In addition to that the non-linear model in Fig. 8.3 appears to fit the observations better, we observe that this non-linear model has varying slopes at different values of the explanatory variable (expr or X). The function is steeper at low values of expr, while it becomes flatter as expr increases. This implies that the slope (measured by the tangent or the derivative of the function) is larger when expr is low. Since the slope in a regression measures the effect of

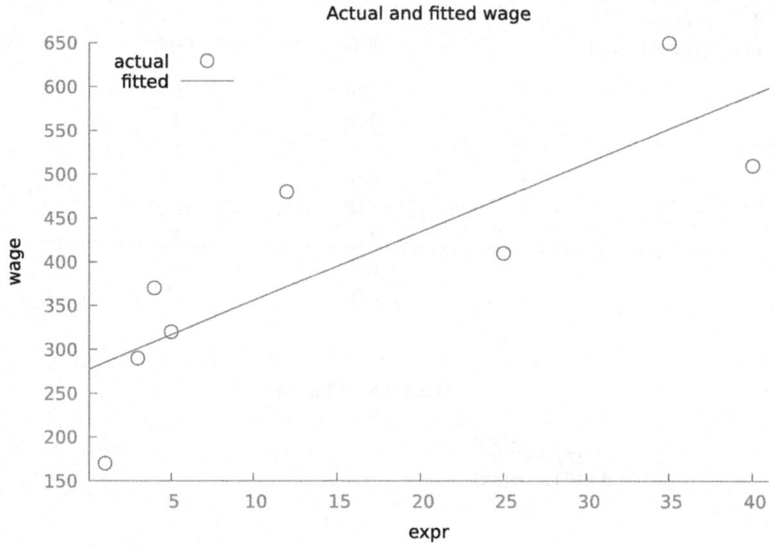

Fig. 8.1 Wage, experience, and regression line

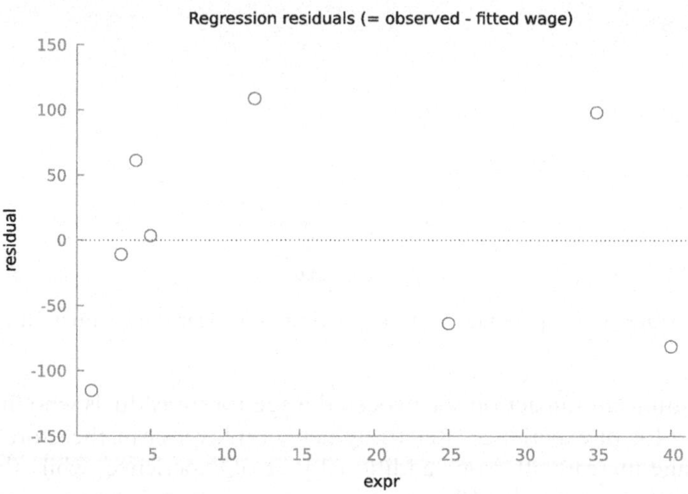

Fig. 8.2 Residuals and experience

the explanatory variable on the dependent variable, it means that such a non-linear model allows for a model where the effect on the dependent variable (wage in our example) depends on the initial level of the explanatory variable (experience).

Because the effect of experience on wage is larger for low values of experience than for high values, this means that one additional year of experience has

Table 8.2 Wage, experience, and residuals

i	*wage*	*expr*	*uhat*
1	320	5	3.4
2	290	3	−10.9
3	170	1	−115.2
4	480	12	108.5
5	510	40	−81.3
6	410	25	−63.6
7	650	35	97.9
8	370	4	61.2

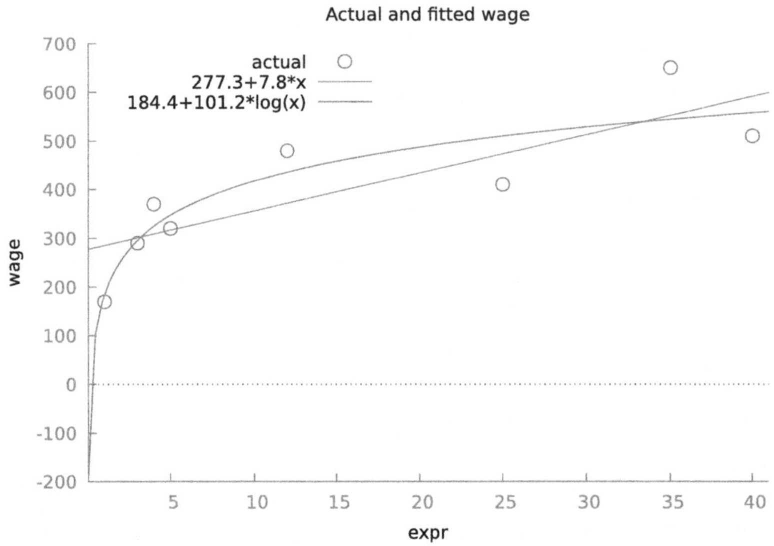

Fig. 8.3 Wage and experience. Linear function and logarithmic function

a more significant impact on the expected wage for individuals who initially do not have a lot of experience. Relatively new employees can therefore expect a higher wage increase after one additional year of experience, while those who already have, for example, 40 years of experience cannot anticipate as high an increase in wages if their experience increases to 41 years. This is because an increase from 1 to 2 years of experience is a doubling or a 100% increase, while an increase from 40 to 41 years of experience is only a 2.5% increase. This non-linear model, therefore, emphasizes the percentage change in the dependent variable rather than a level change.

8.1 Logarithmic Relationships

Often, the logarithm of a variable is taken to obtain a new transformed variable with a distribution closer to the normal distribution. The motivation for using logarithms in the models in this chapter is to account for non-linearity while allowing us to interpret estimated effects with an approximation to percentage growth.

8.1.1 Lin-Log Model

A **lin-log model** is a semi-logarithmic model with the log of the independent variable while the dependent variable is linear. Generally, it takes the form:

$$Y_i = \beta_0 + \beta_1 \ln X_i + u_i \tag{8.2}$$

This means that we use the logarithm of a variable X_i as the explanatory variable.

In Figure 8.3, you can see the straight regression line $\widehat{\text{wage}}_i = 277.4 + 7.8 \cdot \text{expr}_i$, which we estimated earlier. Additionally, you can observe the non-linear relationship $\widehat{\text{wage}}_i = 184.4 + 101.2 \cdot \ln(\text{expr}_i)$, where $\ln(\text{expr}_i)$ represents the natural **logarithm** of (expr_i). We can easily estimate the model:

$$\text{wage}_i = \beta_0 + \beta_1 \ln(\text{expr}_i) + u_i \tag{8.3}$$

by the method of least squares with wage_i as the dependent variable and $\ln(\text{expr}_i)$ as the explanatory variable. The procedures in R and Gretl are illustrated below. Alternatively, you can perform this manually, as shown in Chapter 8.5.2. This model is often referred to as a **lin-log model** because it uses a linear variable as the dependent variable and the logarithm of a linear variable as the explanatory variable.

We obtain:

Model 2: OLS, using observations 1–8
Dependent variable: wage

	Coefficient	Std. Error	*t*-ratio	*p*-value
const	184.421	48.5745	3.797	0.0090***
l_expr	101.197	19.7273	5.130	0.0022***

Mean dependent var	400.0000	S.D. dependent var	148.0347
Sum squared resid	28,482.51	S.E. of regression	68.89910
R^2	0.814325	Adjusted R^2	0.783379
$F(1, 6)$	26.31457	p-value(F)	0.002157
Log-likelihood	−44.06192	Akaike criterion	92.12385
Schwarz criterion	92.28273	Hannan–Quinn	91.05225

As shown in Fig. 8.4 and Table 8.3, the residuals are smaller when we use the logarithm of experience as the explanatory variable compared to using experience as the explanatory variable. This is also reflected in the R-squared value, which indicates that we explain 81.4% of the variation in wage with the logarithm of experience as the explanatory variable, whereas we only had 67.95% explanatory power in the linear model.

Interpreting $\hat{\beta}_1$

Just like in simple regression with linear variables, we can interpret $\hat{\beta}_1$ as the effect on the dependent variable when the explanatory variable increases by 1. For the estimated model $\widehat{\text{wage}}_i = 184.4 + 101.2 \cdot \ln(\text{expr}_i)$, if $\ln(\text{expr}_i)$ increases by 1 year, we expect the wage to increase by 101.2 Euros. This does not provide much practical utility since we want to say something about how experience (measured in years) affects expected wages.

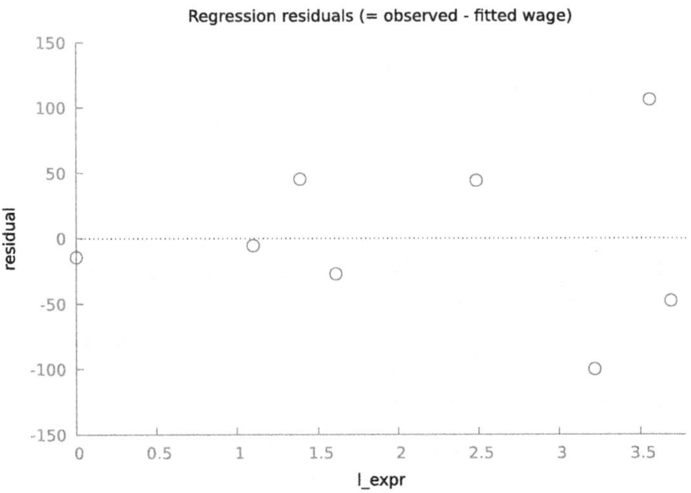

Fig. 8.4 Wage, log experience, and residuals

Table 8.3 Wage, log experience, and residuals

obs	*wage*	*ln (expr)*	*uhat*
1	320	1.61	−27.29
2	290	1.10	−5.60
3	170	0.00	−14.42
4	480	2.48	44.11
5	510	3.69	−47.72
6	410	3.22	−100.16
7	650	3.56	105.79
8	370	1.39	45.29

We can approximate a change in the logarithm of a variable with the relative change in the variable (which can be multiplied by 100% to obtain the percentage change) if the change is small. This allows us to interpret the relationship in a lin-log model as follows: if experience increases by 1% (e.g., from 5 years to 5.05 years), we expect the wage to increase by $\hat{\beta}_1/100 = 1.012$, or 1.012 Euros. See Sect. 8.5.1 for a demonstration of this.

Approximate interpretation in a lin-log model

If we have the estimated model $\hat{Y}_i = \hat{\beta}_0 + \hat{\beta}_1 \ln X_i$, then we have: If X_i increases by 1%, we expect Y_i to change by $\hat{\beta}_1/100$.

A lin-log model is suitable to use when we want to estimate how a percentage change in the explanatory variable affects the level of the dependent variable.

Lin-Log Model in Gretl and R

The only difference between how we estimate $Y_i = \beta_0 + \beta_1 X_i u_i$ and $Y_i = \beta_0 + \beta_1 \ln X_i u_i$ is that we need to use $\ln X_i$ instead of X_i as the explanatory variable. In Gretl, you can select the variable in the main window and then choose `Add >Logs of selected variables`. This will create a new variable named `l_X`, which we can choose as the explanatory variable in the least squares method. Alternatively, you can use the Gretl console to write:

Gretl code 8.1: Logarithm of a Variable

```
l_X = ln(X)
l_expr = ln(expr)
```

In R, you can generate a new variable, such as `lnX`, by using the `log` function or by using the `log()` function directly in the regression (both methods are shown in the code below):

R code 8.1: Logarithm of a Variable and Lin-log Model

```
lnX <- log(X)
Lin-log Model:
lnexpr <- log(expr)
lm(edu ~ lnexpr) or lm(edu ~ log(expr))
```

8.1.2 Log-Lin Model

If we have the logarithm of the dependent variable and the explanatory variable in linear form, we have a **log-lin model** which is another type of a semi-logarithmic model. In general, a simple model in this form is given by:

$$\ln Y_i = \beta_0 + \beta_1 X_i + u_i \tag{8.4}$$

Since we interpret $\hat{\beta}_1$ as the expected effect on the logarithm of Y_i when X_i increases by 1, we can also use an approximation to say something about the effect of X_i on Y_i, since we are more interested in estimating the effect on Y_i than on $\ln Y_i$.

Approximate Interpretation in a Log-lin Model

If we have the estimated model $\widehat{\ln Y}_i = \hat{\beta}_0 + \hat{\beta}_1 X_i$, then we have: If X_i increases by 1, we expect Y_i to change by $(\hat{\beta}_1 \cdot 100)\%$.

A log-lin model is suitable for measuring how a level change in the explanatory variable affects the percentage change or growth in the dependent variable.

For a log-lin model for wage and experience, it would be:

$$\ln \text{wage}_i = \beta_0 + \beta_1 \text{expr}_i + u_i \tag{8.5}$$

and is estimated to

Model 3: OLS, using observations 1–8
Dependent variable: wage

	Coefficient	Std. Error	t-ratio	p-value
const	277.355	46.9739	5.904	0.0010***
expr	7.84928	2.20066	3.567	0.0118**

Mean dependent var	400.0000	S.D. dependent var	148.0347
Sum squared resid	49,161.57	S.E. of regression	90.51848
R^2	0.679520	Adjusted R^2	0.626107
$F(1, 6)$	12.72194	p-value(F)	0.011830
Log-likelihood	−46.24521	Akaike criterion	96.49042
Schwarz criterion	96.64931	Hannan–Quinn	95.41882

The estimated slope coefficient indicates that if experience increases by one year, we expect the wage to increase by 2.05%.

8.1.3 Log-Log Model

If we have the logarithm of both the dependent variable and the explanatory variable, we have a **log-log model**. In general, a simple model in this case will have the form:

$$\ln Y_i = \beta_0 + \beta_1 \ln X_i + u_i \tag{8.6}$$

Since

Approximate interpretation in a log-log model

If we have the estimated model $\widehat{\ln Y}_i = \hat{\beta}_0 + \hat{\beta}_1 \ln X_i$, then we have: If X_i increases by 1%, we expect Y_i to change by $\hat{\beta}_1$%. This measures the **elasticity**.

A log-log model for wage and experience would be:

$$\ln \text{wage}_i = \beta_0 + \beta_1 \ln \text{expr}_i + u_i \tag{8.7}$$

and is estimated as

Model 4: OLS, using observations 1–8
Dependent variable: l_wage

	Coefficient	Std. Error	t-ratio	p-value
const	5.31395	0.123518	43.02	0.0000***
l_expr	0.286183	0.0501637	5.705	0.0013***

Mean dependent var	5.923604	S.D. dependent var	0.411131
Sum squared resid	0.184171	S.E. of regression	0.175200
R^2	0.844345	Adjusted R^2	0.818403
$F(1, 6)$	32.54686	p-value(F)	0.001255
Log-likelihood	3.733820	Akaike criterion	−3.467640
Schwarz criterion	−3.308757	Hannan–Quinn	−4.539242

The estimate of the slope tells us that if experience increases by 1%, we expect the wage to increase by 0.286%.

8.1.4 *Predicted* Y *when* ln Y *Is the Dependent Variable*

If we have the estimated log-log model:

$$\widehat{\ln Y}_i = \hat{\beta}_0 + \hat{\beta}_1 \ln X_i \tag{8.8}$$

then, for example, if $X_i = 5$, we predict $\ln Y_i$ to be:

$$\widehat{\ln Y_i} = \hat{\beta}_0 + \hat{\beta}_1 \ln 5 = \hat{\beta}_0 + \hat{\beta}_1 1.609$$

since $\ln(5) = 1.609$. And if, from our example, we have:

$$\widehat{\ln \text{wage}_i} = 5.31 + 0.29 \ln \text{expr}_i$$

and $\text{expr}_i = 5$, we get:

$$\widehat{\ln \text{wage}_i} = 5.31 + 0.29 \ln 5 = 5.78$$

Hence, the predicted logarithm of the wage is 5.78. Since $e^{\ln X} = X$, we can predict the wage as follows:

$$e^{\widehat{\ln \text{wage}_i}} = e^{5.78}$$
$$\widehat{\text{wage}_i} = 323759$$

Thus, the log-log model predicts the wage to be 323.76 Euros when experience is 5 years. However, this is not a completely accurate approximation, even though it appears to be a preferred approach in practice (see Bårdsen and Lütkepohl [2011]).

8.2 Polynomial Functions

A commonly used method to account for a non-linear effect in regression analysis is to include a squared term of the explanatory variable. This results in a second-degree polynomial function when considering the dependent variable and the explanatory variable. We get the regression model:

$$Y_i = \beta_0 + \beta_1 X_i + \beta_2 X_i^2 + u_i \tag{8.9}$$

Here, we have a **Multiple Regression Model** where both X_i and the squared X_i are explanatory variables.

In Gretl, we can click on the variable we want to square in the main window and then choose `Add > Squares of selected variables` to generate a squared variable, which will be called `sq_X`. Alternatively, we can generate the square of variable `X` by typing `sq_x = X^2` in the console. In R, we can generate the square of variable `X` by typing `sq_X <- X^2`.

In our example, if we use squared experience and experience in such a model to explain wages, we have:

$$\text{wage}_i = \beta_0 + \beta_1 \text{expr}_i + \beta_2 \text{expr}_i^2 + u_i \tag{8.10}$$

This is estimated as:

Model 5: OLS, using observations 1–8
Dependent variable: wage

	Coefficient	Std. Error	*t*-ratio	*p*-value
const	230.556	63.3587	3.639	0.0149**
expr	18.9828	10.5117	1.806	0.1308
sq_expr	−0.279093	0.257830	−1.082	0.3285

Mean dependent var	400.0000	S.D. dependent var	148.0347
Sum squared resid	39,827.97	S.E. of regression	89.25018
R^2	0.740365	Adjusted R^2	0.636511
$F(2, 5)$	7.128911	*p*-value(F)	0.034348
Log-likelihood	−45.40304	Akaike criterion	96.80608
Schwarz criterion	97.04441	Hannan–Quinn	95.19868

So,

$$\widehat{\text{wage}}_i = \hat{\beta}_0 + \hat{\beta}_1 \text{expr}_i + \hat{\beta}_2 \text{expr}_i^2$$
$$\widehat{\text{wage}}_i = 230.56 + 18.98\text{expr}_i - 0.28\text{expr}_i^2$$

The estimated slope coefficients $\hat{\beta}_1 = 18.98$ and $\hat{\beta}_2 = -0.28$ indicate that experience has a positive effect on wages, but squared experience has a negative effect on wages. This is often common when adding a squared explanatory variable, and implies a non-linear effect which is decreasing as the value of the explanatory variable increases. Since the square of a variable is much larger than the level of the variable for high values, adding a squared term "penalizes" the effect of the explanatory variable for high values. If $\text{expr}_i = 2$, then $\text{expr}_i^2 = 4$, whereas if $\text{expr}_i = 40$, then $\text{expr}_i^2 = 1600$. We observe this by taking the derivative of $\widehat{\text{wage}}_i$ with respect to expr_i to find the estimated effect of experience on wages:

$$\frac{\partial \widehat{\text{wage}}_i}{\partial \text{expr}_i} = 18.98 - 2 \cdot 0.28\text{expr}_i = 18.98 - 0.56\text{expr}_i$$

If $\text{expr}_i = 0$, the expected effect of an additional year of experience on wages is 18.98. However, the effect decreases as the initial level of experience increases.

$$\frac{\partial \widehat{\text{wage}}_i}{\partial \text{expr}_i}|(\text{expr}_i = 0) = 18.98 - 0.56 \cdot 0 = 18.98$$
$$\frac{\partial \widehat{\text{wage}}_i}{\partial \text{expr}_i}|(\text{expr}_i = 5) = 18.98 - 0.56 \cdot 5 = 16.18$$
$$\frac{\partial \widehat{\text{wage}}_i}{\partial \text{expr}_i}|(\text{expr}_i = 10) = 18.98 - 0.56 \cdot 10 = 13.38$$

$$\frac{\partial \widehat{\text{wage}}_i}{\partial \text{expr}_i}|(\text{expr}_i = 20) = 18.98 - 0.56 \cdot 20 = 7.78$$

The effect of an additional year of experience on wages is lower for individuals with higher initial experience levels. This means that the effect depends on the level of the explanatory variable, indicating a non-linear effect.

In the linear simple regression model at the beginning of this chapter, we found an effect of 7.85, which was independent of the level of experience. However, the non-linear polynomial regression model we have estimated now shows that the effect is greater than this for low levels of experience, but smaller for high levels. Therefore, 7.85 can be considered an expected effect and is less precise. We can also observe this in Fig. 8.5 since the polynomial regression model is steepest for low experience values and gradually becomes flatter, eventually having a negative slope. This means that the effect of an additional year of experience will be zero where the function has its peak (where the derivative of the second-degree function is zero). For experience values higher than this, the effect will be negative, as shown in the partial derivatives above.

$$\frac{\partial \widehat{\text{wage}}_i}{\partial \text{expr}_i}|(\text{expr}_i = 33.89) = 18.98 - 0.56 \cdot 33.89 = 0.00$$

$$\frac{\partial \widehat{\text{wage}}_i}{\partial \text{expr}_i}|(\text{expr}_i = 40) = 18.98 - 0.56 \cdot 40 = -3.42$$

According to the model, an increase in experience from 40 to 41 years results in a decrease in wages by 3.42 Euros.

To improve the fit and explanatory power, we can add more terms to create a polynomial regression of degree m:

$$Y_i = \beta_0 + \beta_1 X_i + \beta_2 X_i^2 + \beta_3 X_i^3 + \cdots + \beta_m X_i^m + u_i \tag{8.11}$$

Just as when adding more explanatory variables in a multiple regression model, adding more non-linear terms in a polynomial regression reduces degrees of freedom. Therefore, it is essential not to include more terms than needed, and we may consider this by using adjusted R-squared to compare models.

If we include cubic experience (expr_i^3) in addition to squared experience in the model, we get a third-degree polynomial:

$$\widehat{\text{wage}}_i = \hat{\beta}_0 + \hat{\beta}_1 \text{expr}_i + \hat{\beta}_2 \text{expr}_i^2 + \hat{\beta}_3 \text{expr}_i^3$$
$$\widehat{\text{wage}}_i = 181.94 + 37.49\text{expr}_i - 1.46\text{expr}_i^2 + 0.019\text{expr}_i^3$$

To generate a cubic variable from X in Gretl, go to `Add > Define new variable` and write `cu_X = X^3`. In R, you can do this by writing `cu_X <- X^3`. The estimated model with experience,

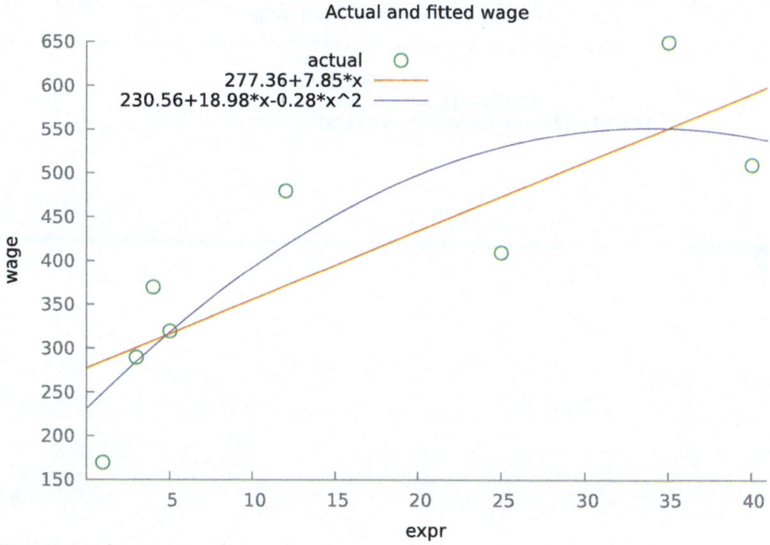

Fig. 8.5 Wage and experience. Linear function and second-degree polynomial function

squared experience, and cubic experience as explanatory variables would then be:

Model 6: OLS, using observations 1–8
Dependent variable: wage

	Coefficient	Std. Error	*t*-ratio	*p*-value
const	181.935	93.3872	1.948	0.1232
expr	37.4877	27.3049	1.373	0.2417
sq_expr	−1.46473	1.62330	−0.9023	0.4179
cu_expr	0.0192655	0.0260088	0.7407	0.5000

Mean dependent var	400.0000	S.D. dependent var	148.0347
Sum squared resid	35,023.78	S.E. of regression	93.57321
R^2	0.771683	Adjusted R^2	0.600446
$F(3, 4)$	4.506509	*p*-value(*F*)	0.089951
Log-likelihood	−44.88887	Akaike criterion	97.77774
Schwarz criterion	98.09551	Hannan–Quinn	95.63454

As we can see in Fig. 8.6, the cubic function might fit the observations even better than the quadratic function. However, the adjusted *R*-squared is lower for the third-degree function compared to the quadratic function. This means that the increased explanatory power gained by adding a cubic term is not significant enough when considering the loss of degrees of freedom in the regression model.

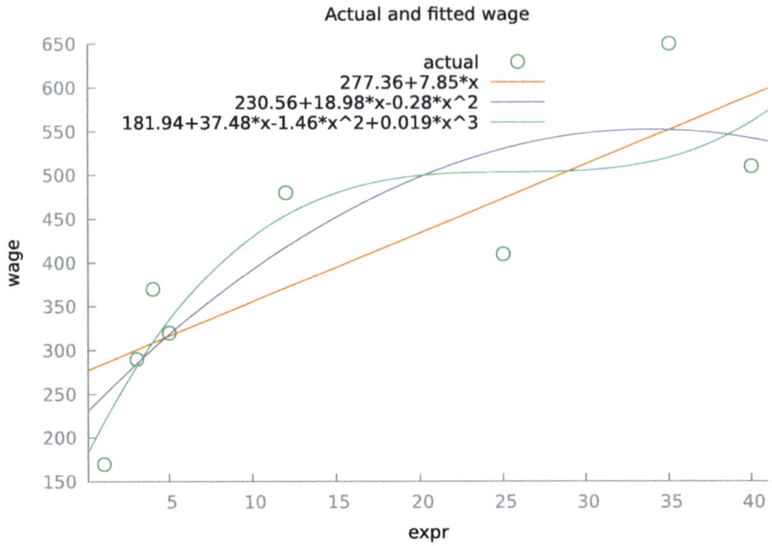

Fig. 8.6 Wage and experience. Linear function, quadratic function, and cubic function

8.3 Inverse Relationships

An alternative to using the logarithm of the explanatory variable is to use its inverse, $1/X_i$. In this case, we have the non-linear model:

$$Y_i = \beta_0 + \beta_1 \frac{1}{X_i} + u_i \tag{8.12}$$

An inverse relationship like this is often used when we have a non-linear negative relationship between two variables.

We can interpret $\hat{\beta}_1$ in an inverse relationship as the expected change in Y_i when $\frac{1}{X_i}$ increases by 1. Since this doesn't tell us much about the relationship between X_i and Y_i, we can, if the change in X_i is small, use the derivative of Y_i with respect to X_i. We then have (given the rules of differentiation—see Sect. 3.2.2):

$$\frac{\partial \hat{Y}_i}{\partial X_i} = -\frac{\hat{\beta}_1}{X_i^2}$$

If we have the wage Y_i for the eight individuals in our example and the number of sick days the employees have had in a year, as shown in Table 8.4, we can use an inverse function to estimate the relationship between sick days and wage, and any potential effect of the number of sick days on wage.

Table 8.4 Wage and number of sick days

i	*1*	*2*	*3*	*4*	*5*	*6*	*7*	*8*
wage	320	290	170	480	510	410	650	370
sickdays	8	8	8	3	2	2	1	2

We then estimate the model:

$$\text{wage}_i = \beta_0 + \beta_1 \frac{1}{\text{sickdays}_i} + u_i \tag{8.13}$$

using the ordinary least squares method. This can be done in Gretl or R by creating a new variable, for example, called `sickdaysinv` In Gretl, you can go to `Add` > `Define new variable...` and enter `sickdaysinv` = `1/sickdays`, or you can do the same in the console window. In R, you can create the new variable by entering `sickdaysinv <- 1/sickdays`

Gretl code 8.2: Creating the inverse variable

```
sickdaysinv = 1/sickdays
```

R code 8.2: Creating the inverse variable

```
sickdaysinv <- 1/sickdays
```

By estimating the model, we obtain the following:

Model 7: OLS, using observations 1–8
Dependent variable: wage

	Coefficient	Std. Error	*t*-ratio	*p*-value
const	226.259	48.5155	4.664	0.0035***
sickdaysinv	433.224	99.3432	4.361	0.0048***

Mean dependent var	400.0000	S.D. dependent var	148.0347
Sum squared resid	36,790.57	S.E. of regression	78.30557
R^2	0.760166	Adjusted R^2	0.720193
$F(1, 6)$	19.01728	*p*-value(F)	0.004766
Log-likelihood	−45.08573	Akaike criterion	94.17146
Schwarz criterion	94.33034	Hannan–Quinn	93.09986

Here we see that $\hat{\beta}_1 = 433.224$. By using the derivative, we can determine the effect of an additional sick day on wages as $-\frac{\hat{\beta}_1}{X_i^2}$. The effect will, therefore, vary depending on what the value of X is. If $X = 1$, meaning the starting point is one sick day, then the effect of one additional sick day will be

$$\frac{\partial \hat{Y}_i}{\partial X_i} = -\frac{\hat{\beta}_1}{X_i^2} = -\frac{433.224}{1^2} = -433.224$$

This means that an extra sick day is expected to result in a wage decrease of 433.22 Euros. If the starting point is 8 sick days, then the expected change in wages for an additional sick day will be

$$\frac{\partial \hat{Y}_i}{\partial X_i} = -\frac{\hat{\beta}_1}{X_i^2} = -\frac{433.224}{8^2} = -6.769$$

In other words, it corresponds to an expected wage decrease of 6.78 Euros. An increase from 8 to 9 sick days is an increase of 12.5%, while an increase from 1 to 2 sick days is an increase of 100%. This non-linear model takes into account the relative change in the explanatory variable to estimate the effect on the dependent variable. If a person initially has many sick days, an additional sick day will not result in a large change in wages. However, if a person has few sick days, an additional sick day will have a greater effect on wages.

8.4 Which Non-Linear Model Should We Choose?

To compare the explanatory power of different models, we often use R-squared or adjusted R-squared for models with different amount of explanatory variables. This can also be done for non-linear models, but only if the dependent variable is the same. For example, from the estimated lin-log model above, 81.4% of the variation in wages is explained by the model. Compared to the lin-lin model, this is better since the lin-lin model explained only 67.95% of the wage variation.

The estimated log-lin model has an R-squared of 84.4%. However, this tells us that the model explains 60.3% of the variation in the *logarithm of wages*. R-squared is, therefore, a measure of something different than in the lin-lin and lin-log models. However, we can compare R-squared in the log-lin model and the log-log model to determine which of these models best explains the variation in $\ln Y_i$.

Comparing Models Using R-squared

We can only compare models using R-squared and adjusted R-squared if the dependent variable is the same in the different models. The dependent variable cannot have different transformations in the models.

We can therefore look at R-squared and adjusted R-squared to evaluate how good a model is, but we cannot always use this to compare two or more models. An assessment of what is best suited theoretically, given what we want to investigate, may be relevant. In addition, we can look graphically at how the non-linear model appears to explain the relationship between two variables.

If we want to estimate an elasticity, it may be reasonable to use a log-log relationship since it provides an estimate of this. If it is appropriate to discuss percentage change or growth in the dependent variable and level change in the explanatory variable, a log-lin model is most suitable, while a lin-log model is better suited if we have a dependent variable that changes in levels while we have a percentage change in the explanatory variable.

Inverse functions often work well when there are negative relationships, while polynomial models can be suitable for estimating non-linear relationships. However, the interpretations of estimated effects are not as straightforward as the approximate interpretations of estimated parameters in logarithmic functions. Therefore, there is a trade-off between having a model which is easy to use for interpreting effects and emphasizing the best possible explanatory power. If the adjusted R-squared for a model with a squared term is higher than for a lin-log model, we should also consider the purpose of the model.

Also, remember that it is not possible to take the logarithm of numbers that are equal to or less than 0, and the logarithmic function approaches negative infinity as it approaches zero. If a variable has negative values, we cannot directly take the logarithm of it. One way to bypass this (if theoretically justifiable) is to transform it first, for example, by adding 100 or another number that "shifts" the variable to be positive in the sample, and then take the logarithm of it. But remember that a percentage change from 4 to 5 is not the same as a percentage change from 104 to 105. When interpreting the estimated parameters, we should take this into account.

Negative Values

Remember that it is not possible to take the logarithm of negative values or zero. If we want to take the logarithm of a variable that contains values less than or equal to zero, we can, for example, transform the variable first.

8.5 Supplementary Material

8.5.1 *Logarithms and Approximation of Percentage Change*

Lin-log Model

Let's assume that X increases from X_1 to X_2, resulting in the values Y_1 and Y_2:

$$Y_2 - Y_1 = (\hat{\beta}_0 + \hat{\beta}_1 \ln X_2) - (\hat{\beta}_0 + \hat{\beta}_1 \ln X_1)$$
$$Y_2 - Y_1 = \hat{\beta}_1(\ln X_2 - \ln X_1)$$

We can approximate $\ln X_2 - \ln X_1$ with the relative change, which gives:

$$Y_2 - Y_1 = \hat{\beta}_1 \frac{X_2 - X_1}{X_1}$$

If $\frac{X_2 - X_1}{X_1} = 1\%$ (X increases by 1

$$Y_2 - Y_1 = \hat{\beta}_1 \cdot 0.01$$
$$Y_2 - Y_1 = \hat{\beta}_1 / 100$$

So, if X increases by 1%, we expect Y to increase by $\hat{\beta}_1/100$.

Log-lin Model

Assuming that X increases from X_1 to X_2, resulting in the values Y_1 and Y_2:

$$\ln Y_2 - \ln Y_1 = (\hat{\beta}_0 + \hat{\beta}_1 X_2) - (\hat{\beta}_0 + \hat{\beta}_1 X_1)$$
$$\ln Y_2 - \ln Y_1 = \hat{\beta}_1(X_2 - X_1)$$

We can approximate $\ln Y_2 - \ln Y_1$ with the relative change, which gives:

$$\frac{Y_2 - Y_1}{Y_1} = \hat{\beta}_1(X_2 - X_1)$$

If $X_2 - X_1 = 1$, we get:

$$\frac{Y_2 - Y_1}{Y_1} = \hat{\beta}_1$$
$$\frac{Y_2 - Y_1}{Y_1} \cdot 100\% = \hat{\beta}_1 \cdot 100\%$$

So, if X increases by 1, we expect Y to increase by $\hat{\beta}_1 \cdot 100\%$.

Log-log Model

Assuming that X increases from X_1 to X_2, resulting in the values Y_1 and Y_2:

$$\ln Y_2 - \ln Y_1 = (\hat{\beta}_0 + \hat{\beta}_1 \ln X_2) - (\hat{\beta}_0 + \hat{\beta}_1 \ln X_1)$$
$$\ln Y_2 - \ln Y_1 = \hat{\beta}_1(\ln X_2 - \ln X_1)$$

We can approximate $\ln Y_2 - \ln Y_1$ and $\ln X_2 - \ln X_1$ with relative changes, which gives:

$$\frac{Y_2 - Y_1}{Y_1} = \hat{\beta}_1 \frac{X_2 - X_1}{X_1}$$

If $\frac{X_2 - X_1}{X_1} = 1\%$ (X increases by 1%), we get:

$$\frac{Y_2 - Y_1}{Y_1} = \hat{\beta}_1 \cdot 0.01$$
$$\frac{Y_2 - Y_1}{Y_1} = \hat{\beta}_1\%$$

So, if X increases by 1%, we expect Y to increase by $\hat{\beta}_1\%$.

8.5.2 *Lin-Log Regression by Hand*

If we have wage as the dependent variable and the logarithm of experience as the independent variable, we have the lin-log model:

$$Y_i = \beta_0 + \beta_1 \ln X_i + u_i$$

as shown in (8.2), where Y_i is wage and X_i is experience. We need to find:

$$\hat{\beta}_1 = \frac{\sum(Y_i - \bar{Y})(\ln X_i - \overline{\ln X})}{\sum(\ln X_i - \overline{\ln X})^2}$$
$$\hat{\beta}_0 = \bar{Y} - \hat{\beta}_1 \overline{\ln X}$$

So, we have replaced X_i with $\ln X_i$ as we did in (5.5) and (5.6). We need to calculate $\sum(Y_i - \bar{Y})(\ln X_i - \overline{\ln X})$ and $\sum(\ln X_i - \overline{\ln X})^2$, similar to simple regression, as shown in Table 5.1. First, we find that $\overline{\ln X} = \sum_{i=1}^{8} \overline{\ln X}/n = 17.042/8 = 2.130$. The rest of the calculations are shown in Table 8.5.

We then have:

$$\hat{\beta}_1 = \frac{\sum(Y_i - \bar{Y})(\ln X_i - \overline{\ln X})}{\sum(\ln X_i - \overline{\ln X})^2} = \frac{1234.404}{12.198} = 101.197$$
$$\hat{\beta}_0 = \bar{Y} - \hat{\beta}_1 \overline{\ln X} = 400 - 101.197 \cdot 2.130 = 184.42$$

Table 8.5 Calculations for estimating coefficients in a lin-log model

$\ln X$	Y	$\ln X - \overline{\ln X}$	$Y - \bar{Y}$	$(\ln X - \overline{\ln X})(Y - \bar{Y})$	$(\ln X - \overline{\ln X})^2$
1.609	320	−0.521	−80	41.669	0.271
1.099	290	−1.032	−110	113.485	1.064
0.000	170	−2.130	−230	489.968	4.538
2.485	480	0.355	80	28.369	0.126
3.689	510	1.559	110	171.444	2.429
3.219	410	1.089	10	10.886	1.185
3.555	650	1.425	250	356.263	2.031
1.386	370	−0.744	−30	22.320	0.554
17.042	3200	0.000	0	1234.404	12.198

Reference

Bårdsen, G., & Lütkepohl, H. (2011). Forecasting levels of log variables in vector autoregressions. *International Journal of Forecasting, 27*(4), 1108–1115.

CHAPTER 9

Time Series Data

When having time series data, there are many challenges that we can meet. This chapter covers the main issues, and relevant tests, for stationarity, autocorrelation, trends, seasonal effects, and parameter stability. We also use different time series models for forecasting, including autoregressive models and manipulating time series data, and look at cointegration to estimate long-run effects.

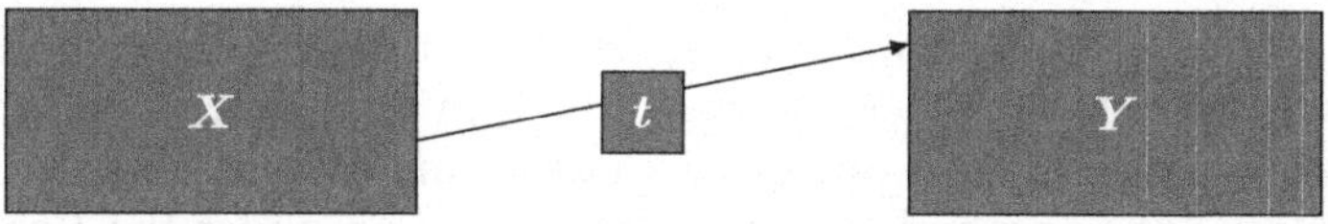

In this chapter, we will explore some challenges associated with using time series data, and introduce some fundamental models for analyzing time series data. However, for more comprehensive analyses of time series variables, it may be necessary to move on to other models beyond those presented here. This chapter serves as an introduction to time series analysis.

Supplementary Information The online version contains supplementary material available at https://doi.org/10.1007/978-3-031-53142-2_9.

B. Karlsen Kivedal, *Applied Statistics and Econometrics*,
https://doi.org/10.1007/978-3-031-53142-2_9

9.1 Time Series Variables and Regressions

In the simple regression model (see Chapter 5), we had

$$Y_i = \beta_0 + \beta_1 X_i + u_i$$

where Y_i was the dependent variable, and X_i was the independent variable.

This is a model for cross-sectional data (data collected at a single point in time) measured, for example, for each individual i. If we have time series data, we use **subscript** t since we have data for period t and not for an individual or entity i. We then have

$$Y_t = \beta_0 + \beta_1 X_t + u_t$$

For example, Y_t can be the quantity of a product sold in year t, and X_t is the price of the product in year t.

If we have observed quantity sold Y_t measured in units and price X_t measured in our currency for six years, $t = [1, 2, 3, 4, 5, 6]$, as shown in Table 9.1, we can estimate a regression model to assess the relationship between price and quantity sold. Here, we have a total of 6 periods with observations, which we typically denote as $T = 6$ (not $n = 6$, which we would use for cross-sectional data with observations for, say, 6 firms).

9.1.1 *Time Series in Gretl and R*

In Gretl, when working with time series data, it is important to structure the dataset correctly to take advantage of built-in time series functions. When creating a new dataset, you can specify it as a time series dataset by going to `File > New dataset` and selecting `Time series` under the `Structure of dataset` option. You will also need to specify the number of observations (in our example, $T = 6$). We should also indicate the frequency of the data, whether it is annual, quarterly, monthly, weekly, etc. In our example, it is annual (you can select `annual`). Additionally, you need to specify the starting period, which, in our example, is 2001 (Fig. 9.1).

Table 9.1 Sold quantity and price six years, 2001–2006

Year	t	*Sales*	*Price*
2001	1	4,000	70
2002	2	5,500	68
2003	3	5,000	79
2004	4	6,000	85
2005	5	7,000	91
2006	6	7,500	89

(a) Choosing the structure of the data set

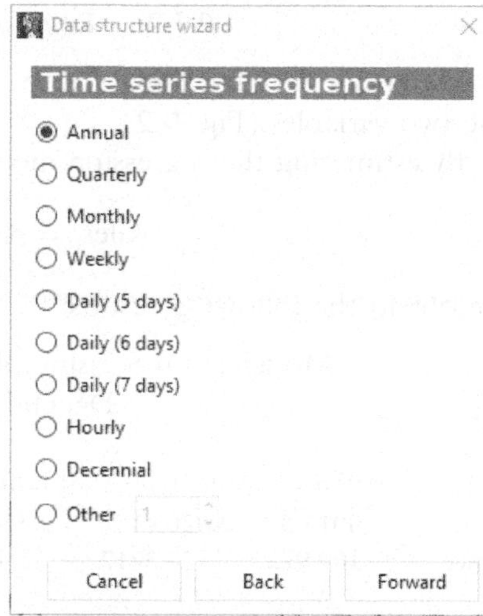

(b) Choosing time series frequency

Fig. 9.1 Time series data in Gretl

In R, you can use the `ts()` function to classify a variable as a time series. You should also specify the starting point and the frequency of the time series data.

R code 9.1: Time series i R

```
sales <- c(4000,5500,5000,6000,7000,7500)
price <- c(70,68,79,85,91,89)

sales <- ts(sales, frequency = 1, start=c(2001))
price <- ts(price, frequency = 1, start=c(2001))

aar <- seq(as.Date("2001-01-01"), length = 6, by = "year")
```

Here, we first create the variables and then classify them as time series variables. This classification allows us to automatically display the time series as lines in a plot. The `frequency = 1` command indicates that there is 1 observation per year since the data is annual. `2001` indicates that the observations start in the year 2001. If we had quarterly data starting in the first quarter, we would use `frequency = 4, start=c(2001,1))`. If we had monthly data starting in May 2007, we would use `frequency = 12, start=c(2007,5))`. The number after the comma (e.g., 1) specifies the quarter or month in which the dataset begins. Finally, we can create a date variable for the year, as shown at the end of the R code.

9.1.2 Regression with Time Series

A scatterplot of sales and price reveals a strong positive correlation between the two variables (Fig. 9.2).

By estimating the regression model:

$$\text{sales}_t = \beta_0 + \beta_1 \text{price}_t + u_t$$

we obtain the following results:

Model 1: OLS, using observations 2001–2006 ($T = 6$)
Dependent variable: sales

	Coefficient	*Std. error*	t-*ratio*	p-*value*
const	−3,001.06	3,020.41	−0.9936	0.3767
price	109.972	37.3717	2.943	0.0423**

Mean dependent var	5,833.333	S.D. dependent var	1,290.994
Sum squared resid	2,633,133	S.E. of regression	811.3466
R^2	0.684024	Adjusted R^2	0.605030
$F(1, 4)$	8.659191	P-value (F)	0.042277
Log-likelihood	−47.48941	Akaike criterion	98.97882
Schwarz criterion	98.56233	Hannan–Quinn	97.31161
$\hat{\rho}$	−0.555494	Durbin–Watson	2.518289

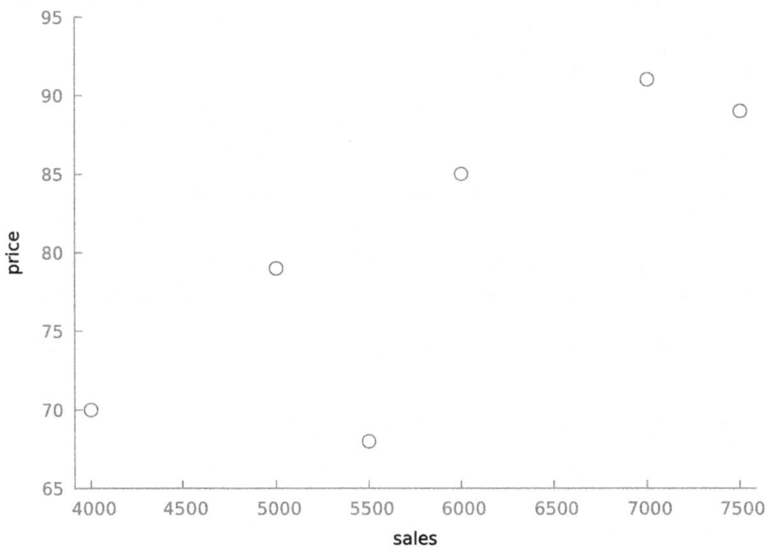

Fig. 9.2 Scatterplot of sales and price

Therefore, we have the estimated model:

$$\widehat{\text{sales}_t} = \hat{\beta}_0 + \hat{\beta}_1 \text{price}_t + u_t$$

With our estimated coefficients, it becomes:

$$\widehat{\text{sales}_t} = -3001.06 + 109.97 \cdot \text{price}_t + u_t$$

Here, we observe that sales are negative when the price is zero in a year. This is outside the observation range and therefore makes little sense to interpret. However, we have estimated that for every one-unit increase in price, we can expect an increase in sales by 110 units. This effect is significant at a 5% significance level. Thus, for this product, we see from the simple regression model that sales increase as the price rises, which is unusual for regular products (known as normal goods).

In our example, we observe a very high R-squared ($R^2 = 0.993$), indicating that the model (with price as the explanatory variable) explains 99.3% of the variation in the quantity sold. It is often the case in regressions with time series data that exhibit growth over time to have a very high R-squared. However, when working with time series data, it is not as useful to examine R-squared as for cross-sectional data. This is because we are not interested in explaining the variation in sales but rather in understanding the evolution of sales over time.

Measuring Explained Variance for Time Series

When evaluating the quality of models with time series variables, it can be problematic to focus on R-squared.

We will explore various measures for assessing time series models in Sect. 9.4.

9.1.3 Development Over Time: Trend

It is more common to examine the evolution of variables over time when using time series data. We can visualize this by creating a time series plot. In Gretl, you can do this by going to `View` > `Graph specified vars` > `Time series plot...` and selecting the variables you want to display in the figure. For quantity sold and price, it would look like Fig. 9.3.

In R, you can create a time series plot using the `plot(variable_name)` function. To display multiple graphs simultaneously, you first need to create a plot with one variable and then add the others.

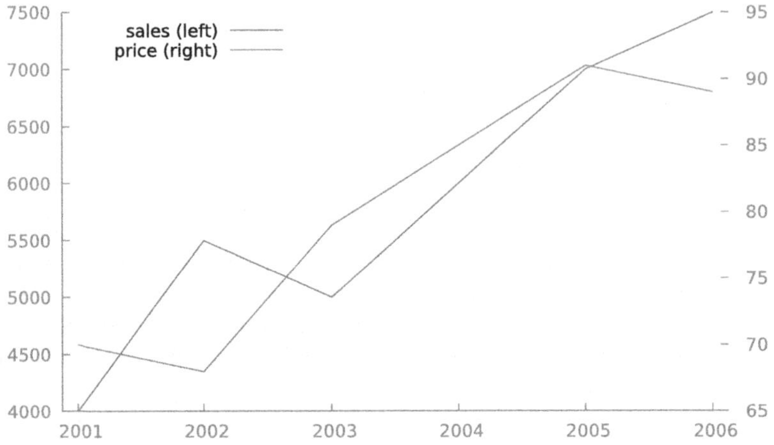

Fig. 9.3 Time series plot of quantity sold and price

R code 9.2: Time series plot for two variables in R

```
plot(sales, col="blue", ylab="sales (blue)")
par(new = TRUE)
plot(price, axes = FALSE, bty = "n",
       xlab = "", ylab = "", col="red")
axis(side=4, at = pretty(range(price)))
mtext("price (red)", side=4, line=3)
```

This produces the graphs shown in Fig. 9.4.

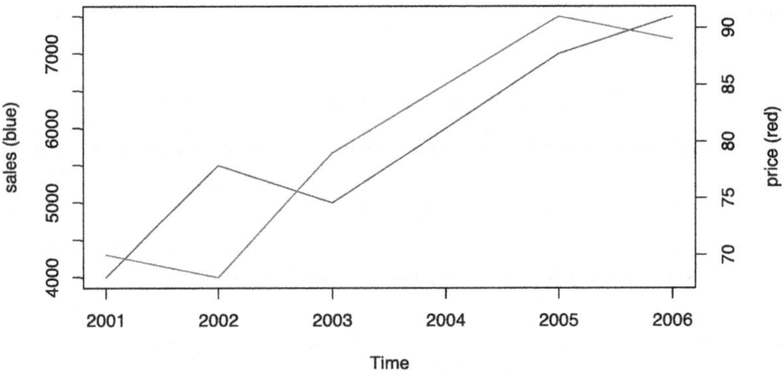

Fig. 9.4 Time series plot of quantity sold and price created in R

The time series graphs show that both quantity sold and price have a positive trend over time. As seen in Fig. 9.3, there is a positive trend in both sales and price over the six years for which we have observations. It is possible that the growth in sales and price is influenced by other factors, such as a general increase in the price level and a general increase in demand for the product we are selling. They may share a common trend.

A variable commonly used in time series analysis is a **trend**. We will denote it as t, which takes the value $t = 1$ for the first data point in the dataset and $t = T$ for the last data point. We already saw this variable in Table 9.1 for our example, taking the values from 1 to 6.

In Gretl, you can create a trend variable by going to `Add > Time trend`, while in R, you can create a trend variable using the code snippet below:

R code 9.3: Trend in R

```
t <- (1:6)
t <- ts(t, frequency = 1, start=c(2001,1))
```

Regression with Trend as an Explanatory Variable

We can add the trend variable as an explanatory variable in the regression model to estimate the model:

$$Y_t = \beta_0 + \beta_1 X_t + \beta_2 t + u_t \tag{9.1}$$

In this model, the estimated value $\hat{\beta}_2$ can be interpreted as the expected increase in sales from one year to the next, given that the price does not change. $\hat{\beta}_1$ can tell us what a price increase of 1 within a given year is expected to have on sales. By estimating the model, we get the following:

Model 2: OLS, using observations 2001–2006 ($T = 6$)
Dependent variable: sales

	Coefficient	*Std. error*	*t-ratio*	*p-value*
const	7,891.37	3,400.11	2.321	0.1030
price	−68.8111	53.3500	−1.290	0.2876
time	991.368	276.872	3.581	0.0373**

Mean dependent var	5,833.333	S.D. dependent var	1290.994
Sum squared resid	499,307.8	S.E. of regression	407.9656
R^2	0.940083	Adjusted R^2	0.900138
$F(2, 3)$	23.53466	P-value (F)	0.014666
Log-likelihood	−42.50129	Akaike criterion	91.00257
Schwarz criterion	90.37785	Hannan–Quinn	88.50176
$\hat{\rho}$	−0.530054	Durbin–Watson	2.859979

Here we can observe a negative effect of an increased price on sales, assuming there is no trend. This aligns more with normal goods, although the effect is not significant. The estimated effect of the trend is $\hat{\beta}_2 = 991.4$, indicating that we expect an increase in sales of 991.4 units per year, assuming the price remains unchanged. We have now taken into account both a potential change over time by including the trend variable and the effect of price to make conclusions about the quantity sold.

9.1.4 *Spurious Regression*

As shown in the regression model with quantity sold and price, we obtained a very high R-squared and a significant positive effect of price on the quantity sold. This could be due to a common trend for the two variables, but it could also be coincidental, where both variables have grown over time for different reasons. In such cases, the significant effect can be attributed to a random or **spurious relationship**, which means we find a significant effect that we cannot explain or find any background for, for example using economic theory.

A widely used example of this is the correlation between the number of fatal motorway accidents in the USA and the number of tons of fresh lemons imported from Mexico to the USA, as documented in Johnson (2008). As shown in Fig. 9.5, there is a negative correlation (Table 9.2).

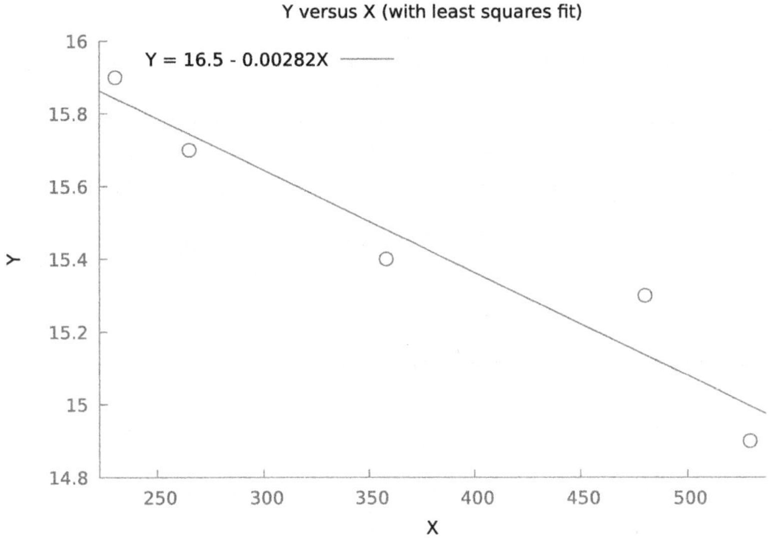

Fig. 9.5 Scatterplot and regression line for the total motorway accident fatality rate in the USA (Y_t) and the number of tons of fresh lemons imported from Mexico to the USA (X_t)

Table 9.2 Total motorway accident fatality rate in the USA (Y_t) and the number of tons of fresh lemons imported from Mexico to the USA (X_t)

Year	Y	X
1996	15.9	230
1997	15.7	265
1998	15.4	358
1999	15.3	480
2000	14.9	530

By estimating the regression model

$$Y_t = \beta_0 + \beta_1 X_t + u_t$$

where X_t represents lemons and Y_t represents the fatality rate, we get the following results:

Model 3: OLS, using observations 1996–2000 ($T = 5$)
Dependent variable: Y

	Coefficient	*Std. error*	t-*ratio*	p-*value*
const	16.4909	0.188028	87.70	0.0000***
X	−0.00282048	0.000481463	−5.858	0.0099***

Mean dependent var	15.44000	S.D. dependent var	0.384708
Sum squared resid	0.047591	S.E. of regression	0.125951
R^2	0.919609	Adjusted R^2	0.892812
$F(1, 3)$	34.31777	P-value (F)	0.009918
Log-likelihood	4.541667	Akaike criterion	−5.083334
Schwarz criterion	−5.864458	Hannan–Quinn	−7.179794
$\hat{\rho}$	−0.726156	Durbin–Watson	2.906665

The R-squared is 0.9196, which means that 91.96% of the variation in the US fatality rate can be explained by lemon imports from Mexico. Additionally, the relationship is highly significant. However, this is spurious regression, and we find a significant relationship only because there happened to be both a decrease in motorway accidents in the USA and an increase in lemon imports from Mexico during the period, as shown in Fig. 9.6.

By including the trend in the regression model, we can control for any potential common trend when trying to estimate the effect of lemon imports on motorway accidents. In this case, we obtain the following results:

Model 4: OLS, using observations 1996–2000 ($T = 5$)

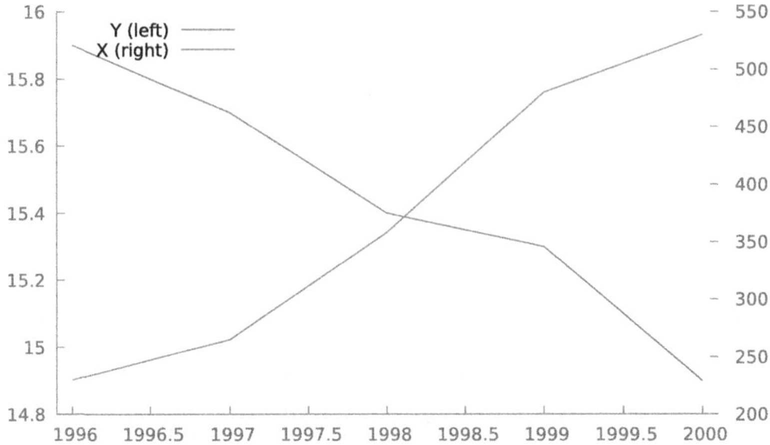

Fig. 9.6 Time series of the total motorway accident fatality rate in the USA (Y_t) and the number of tons of fresh lemons imported from Mexico to the USA (X_t)

Dependent variable: Y

	Coefficient	*Std. error*	t-*ratio*	p-*value*
const	15.9958	0.242329	66.01	0.0002***
X	0.00128186	0.00177572	0.7219	0.5454
time	−0.344472	0.146897	−2.345	0.1437

Mean dependent var	15.44000	S.D. dependent var	0.384708
Sum squared resid	0.012693	S.E. of regression	0.079664
R^2	0.978559	Adjusted R^2	0.957119
$F(2, 2)$	45.64061	P-value (F)	0.021441
Log-likelihood	7.845703	Akaike criterion	−9.691406
Schwarz criterion	−10.86309	Hannan–Quinn	−12.83610
$\hat{\rho}$	−0.863920	Durbin–Watson	2.960412

Here, we can see that there is no significant relationship between lemon imports and motorway accidents. One solution to spurious regression can thus be to include a trend in the model.

Table 9.3 Lagged variables

Year	t	*Sales*	*Sales_1*	*Sales_2*
2001	1	4,000		
2002	2	5,500	4,000	
2003	3	5,000	5,500	4,000
2004	4	6,000	5,000	5,500
2005	5	7,000	6,000	5,000
2006	6	7,500	7,000	6,000

9.1.5 Transforming Time Series Variables

Lagged Variables

Since we have observations for different time periods, we can create a variable that shows the value of one or more periods back in time. These variables will contain only the values of the original variable but with a shift in the observation time. In Table 9.3, we have created lagged variables one and two periods back in time for our example with sales.

So, we have the variable $sales\_ 1_t$ with values $sales_{t-1}$ and the variable $sales\_ 2_t$ with values $sales_{t-2}$. We can treat these as new variables in, for example, a regression analysis where we can include lagged variables as explanatory variables.

Lagged Variables in Gretl and R

In Gretl, we can create lagged variables by selecting `Add > Lags of selected variables` for the variable we desire. Then, we choose how many periods back in time we want lagged variables for. If we select 2, we will get variables as shown in Table 9.3.

R code 9.4: Lagged variables in R

```
install.packages("dplyr")  #only the first time
library(dplyr)
sales1 <- lag(sales)
sales2 <- lag(sales, 2)

price1 <- lag(price)
price2 <- lag(price, 2)
```

Table 9.4 Sold quantity and price. Level and change

Year	t	*Sales*	*Price*	*dsales*	*dprice*
2001	1	4,000	70		
2002	2	5,500	67	1,500	−3
2003	3	5,000	79	−500	12
2004	4	6,000	85	1,000	6
2005	5	7,000	91	1,000	6
2006	6	7,500	89	500	−2

Differenced Variables

To avoid variables that grow or decline over time, we can calculate changes in the variable. This is often denoted with the Greek uppercase delta:

$$\Delta Y_t = Y_t - Y_{t-1}$$

We cannot use the first observation since it serves as the reference point for calculating changes (Table 9.4).

where $\text{dsales}_t = \text{sales}_t - \text{sales}_{t-1}$. These are referred to as **differenced variables** or the **first difference** of a variable. Since the sales increase from $\text{sales}_1 = 4000$ to $\text{sales}_2 = 5500$ from period 1 to 2, we have $\Delta\text{sales}_2 = \text{sales}_2 - \text{sales}_1 = 5500 - 4000 = 1500$.

In Fig. 9.7, we observe changes in the quantity sold and in price from the previous year for the years 2002–2006. These time series do not appear to exhibit positive or negative growth over time, which can resolve the issue of spurious regression. If we estimate the model

$$\Delta\text{sales}_t = \beta_0 + \beta_1 \Delta\text{price}_t + u_t \tag{9.2}$$

we obtain

Model 5: OLS, using observations 2002–2006 ($T = 5$)
Dependent variable: d_sales

	Coefficient	*Std. error*	*t-ratio*	*p-value*
const	1,018.63	380.881	2.674	0.0754*
d_price	−83.8509	60.0726	−1.396	0.2571

Mean dependent var	700.0000	S.D. dependent var	758.2875
Sum squared resid	1,394,410	S.E. of regression	681.7649
R^2	0.393735	Adjusted R^2	0.191646
$F(1, 3)$	1.948330	P-value (F)	0.257133
Log-likelihood	−38.44105	Akaike criterion	80.88211
Schwarz criterion	80.10098	Hannan–Quinn	78.78565
$\hat{\rho}$	−0.621327	Durbin–Watson	2.414489

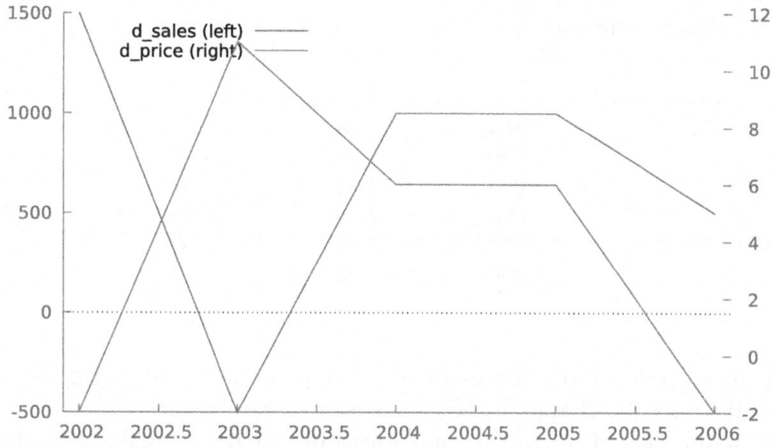

Fig. 9.7 First difference of sales and price

Hence, we do not have a positive significant relationship between price and quantity, but a non-significant negative relationship, just like in the model where we included a trend. Note that here we only have $T = 5$ observations since the observations for Δsales and Δprice start in 2002.

Creating Differentiated Variables in Gretl and R

In Gretl, you can create differentiated variables by going to `Add > First differences of selected variables`. This will generate variables like d_sales_t and d_price_t.

To create differentiated variables in R, you can generate a variable that represents the difference between a variable and its lagged value for one period. For example: $\text{dsales}_t = \text{sales}_t - \text{sales}_{t-1}$.

R code 9.5: Creating Differentiated Variables in R

```
dsales <- sales - lag(sales)
dprice <- price - lag(price)
```

Calculating Growth

Percentage growth in a variable for period t is measured as

$$X_{\text{growth}} = \frac{X_t - X_{t-1}}{X_{t-1}} \cdot 100$$

which represents the change from the previous period $(t - 1)$ to period t divided by the level of the previous period. To calculate growth, you can divide a differenced variable by a lagged variable if you have already computed them.

Table 9.5 Growth in sales

Year	t	*sales*	*lsales*	*salesgrowth*	*dlsales*
2001	1	4,000	8.29		
2002	2	5,500	8.61	37.50	31.85
2003	3	5,000	8.52	–9.09	–9.53
2004	4	6,000	8.70	20.00	18.23
2005	5	7,000	8.85	16.67	15.42
2006	6	7,500	8.92	7.14	6.90

The fraction shown above represents the change in rates, but you can multiply it by 100 to get the change in percentages.

Often, the change in the natural logarithm of the variable is used as an approximation of this (see Sect. 8.5.2), which can be used when there is a low percentage change.

$$X_{\text{growth}} = \frac{X_t - X_{t-1}}{X_{t-1}} \cdot 100 \approx (\ln X_t - \ln X_{t-1}) \cdot 100$$

Table 9.5 shows the variables sales, lsales, salesgrowth, and dlsales, which represent sales from the example we have used, the natural logarithm of sales, growth in percentage as calculated using the growth formula here, and the difference in the logarithm of sales multiplied by 100.

We see from the table that approximating growth by the difference of the log not is very accurate when the growth is high. This approximation is often used when dealing with high frequency financial data where changes are not very large, and may thus often be appropriate to use.

Growth Variables in Gretl and R

In Gretl, you can create growth variables measured as $\frac{X_t - X_{t-1}}{X_{t-1}} \cdot 100$ by going to `Add > Percentage change of selected variables`.

In R, you can use the `lag()` function to calculate growth directly. For example:

R code 9.6: Calculating Growth in R

```
salesgrowth <- (sales-lag(sales))/lag(sales)*100
pricegrowth <- (price-lag(price))/lag(price)*100
```

9.2 Autocorrelation and Stationarity

As we saw in the previous subsection, there are challenges when dealing with time series variables. In particular, dealing with variables that grow over time can be problematic, as we may encounter issues with spurious correlation. Even though we observed from the graphs in the examples that the variables had positive or negative growth over time, it is essential to test this rigorously. We can use hypothesis tests to examine how a variable evolves over time.

Autocorrelation is present when a variable is correlated with its past values, meaning that Y_t depends on Y_{t-1}. This is called **first-order autocorrelation**. Additionally, we can have situations where Y_t depends on a value of itself p periods back in time, Y_{t-p}, which is autocorrelation of order p. For instance, when dealing with quarterly data, it might be relevant to consider $p = 4$ because the period $t - 4$ corresponds to the same quarter and season as period t. In particular, we should avoid autocorrelated errors to ensure that our model is well-specified.

If there is autocorrelation, the assumption that we have a random sample (see Assumption 2 in Sect. 5.7) is violated because one period is correlated with the period or other periods before systematically. As mentioned earlier in this chapter, this can lead to spurious regression, which is something we should be cautious about. However, autocorrelation can provide information about the development of a variable over time, which can be useful for making **forecasts**.

9.2.1 *Autoregressive Model*

An autoregressive model of the first order (an AR(1) model) is a model in which a variable depends on its value one period earlier. In other words, we can express it as follows:

$$X_t = \beta_0 + \beta_1 X_{t-1} + u_t \tag{9.3}$$

The value of β_1 indicates to what extent the value in the previous period affects the current value and whether there is a positive or negative relationship from period to period. Additionally, we can determine whether X_t is **stationary** or **non-stationary**. If $-1 < \beta_1 < 1$, then X_t is stationary, and if $\beta_1 \geq 1$ or $\beta_1 \leq -1$, then X_t is non-stationary. A stationary variable is a variable that has the same probability distribution over time, meaning that the mean and variance remain constant. Therefore, a stationary variable should fluctuate around a mean value over time, while for a non-stationary variable, changes in the mean or variance can be observed over time.

In Fig. 9.8, a stationary variable Z_t and a non-stationary variable W_t are illustrated. It is evident that W_t grows over time, causing the mean value to increase, whereas Z_t maintains a mean of 0.5 throughout the period.

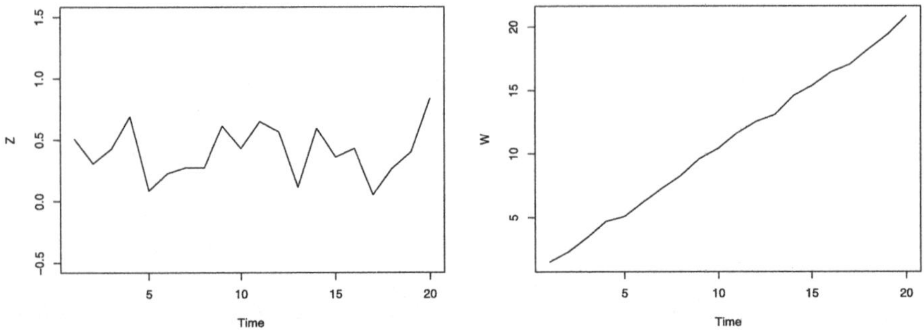

Fig. 9.8 Plot of Z_t and W_t

By estimating AR(1) models for these two variables, we get

$$\hat{Z}_t = \underset{(0.12)}{0.43} - \underset{(0.27)}{0.079}\ Z_{t-1}$$

$$\hat{W}_t = \underset{(0.15)}{0.93} + \underset{(0.013)}{1.01}\ W_{t-1}$$

Since $-1 < \hat{\beta}_1 < 1$ for Z_t and $\hat{\beta}_1 \geq 1$ for W_t, this suggests that Z_t is stationary, while W_t is non-stationary.

Regressions with non-stationary variables can yield estimates that are not unbiased, making it challenging to perform hypothesis tests, such as t-tests. There is also a risk of finding spurious effects, as demonstrated in the previous section. Therefore, we should use stationary variables when employing the regression model. Alternatively, we can use other methods if we have non-stationary variables, which we will consider later.

Non-stationary variables can often be transformed into stationary variables by taking the difference or percentage change of the variable. Thus, we can generate a stationary variable that retains much of the same variation as the non-stationary variable, even if the levels are not the same.

We will examine an example using twelve quarters (two and a half years), $T = 12$, of sales data to analyze the temporal development (Fig. 9.9).

We observe that sales are growing over time, and it is possible that the time series is autocorrelated. We can estimate the model:

$$\text{sales}_t = \beta_0 + \beta_1 \text{sales}_{t-1} + u_t$$

to investigate this. This results in the following:

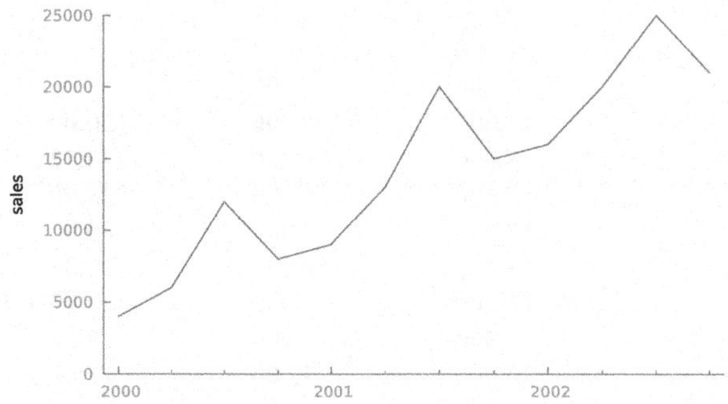

t	sales
1	4,000
2	6,000
3	12,000
4	8,000
5	10,000
6	13,000
7	20,000
8	15,000
9	17,000
10	20,000
11	25,000
12	21,000

Fig. 9.9 Quarterly data for sales first quarter 2000 ($t = 1$) to fourth quarter 2002 ($t = 12$)

Model 6: OLS, using observations 2000:2–2002:4 ($T = 11$)
Dependent variable: sales

	Coefficient	*Std. Error*	t-*ratio*	p-*value*
const	5,347.68	2,852.09	1.875	0.0935*
sales_1	0.721170	0.190409	3.787	0.0043***

Mean dependent var	15,181.82	S.D. dependent var	5,980.271
Sum squared resid	1.38e+08	S.E. of regression	3,914.021
R^2	0.614480	Adjusted R^2	0.571644
$F(1, 9)$	14.34507	P-value (F)	0.004299
Log-likelihood	−105.5002	Akaike criterion	215.0003
Schwarz criterion	215.7961	Hannan–Quinn	214.4987
$\hat{\rho}$	−0.285055	Durbin's h	−1.219322

We observe that $\hat{\beta}_1 = 0.72$. Therefore, the sales variable may be non-stationary. We can test this using the hypothesis test $H_0 : \beta_1 = 1$ versus $H_A : \beta_1 \neq 1$. However, we cannot use the t-distribution as usual because the null hypothesis implies non-stationarity. Thus, we will not focus on conducting formal tests for non-stationarity here, but rather make an assessment by visually inspecting the time series and the estimated value $\hat{\beta}_1$ in an AR(1) model. A formal test will be discussed in Sect. 9.2.5.

Table 9.6 Quarterly data for sales, changes (dsales), and growth (salesgrowth)

Time	*Year*	*Quarter*	*Sales*	*dsales*	*Salesgrowth*
1	2000	1	4,000		
2	2000	2	6,000	2,000	50.00
3	2000	3	12,000	6,000	100.00
4	2000	4	8,000	−4,000	−33.33
5	2001	1	10,000	2,000	25.00
6	2001	2	13,000	3,000	30.00
7	2001	3	20,000	7,000	53.85
8	2001	4	15,000	−5,000	−25.00
9	2002	1	17,000	2,000	13.33
10	2002	2	20,000	3,000	17.65
11	2002	3	25,000	5,000	25.00
12	2002	4	21,000	−4,000	−16.00

To use the regression model to make predictions about sales, we can transform the variable into changes or growth rates in sales (Table 9.6).

By estimating the model:

$$\text{dsales}_t = \beta_0 + \beta_1 \text{dsales}_{t-1} + u_t$$

where $\text{dsales}_t = \text{sales}_t - \text{sales}_{t-1}$, we obtain the following results:

Model 7: OLS, using observations 2000:3–2002:4 ($T = 10$)
Dependent variable: d_sales

	Coefficient	*Std. error*	*t-ratio*	*p-value*
const	2,550.77	1,500.45	1.700	0.1276
d_sales_1	−0.500365	0.352681	−1.419	0.1937

Mean dependent var	1,500.000	S.D. dependent var	4,352.522
Sum squared resid	1.36e+08	S.E. of regression	4,126.515
R^2	0.201027	Adjusted R^2	0.101155
$F(1, 8)$	2.012848	P-value (F)	0.193734
Log-likelihood	−96.32555	Akaike criterion	196.6511
Schwarz criterion	197.2563	Hannan–Quinn	195.9872
$\hat{\rho}$	−0.235588	Durbin's h	NA

Here, $\hat{\beta}_1 = -0.50$. However, as indicated by the p-value, the effect is not significant. A change in sales in the previous quarter does not have a significant relationship with the change in sales in the current quarter.

9.2.2 *Autoregressive Model with Multiple Lags*

We can also include multiple lagged terms in an autoregressive model. This results in an AR(p) model, where X depends on X p periods earlier. In other words, we have the model:

$$X_t = \beta_0 + \beta_1 X_{t-1} + \beta_2 X_{t-2} + \cdots + \beta_p X_{t-p} + u_t \tag{9.4}$$

To determine the number of lagged terms, we can use **information criteria**. These criteria measure prediction error of the model while taking into account the number of lagged terms included in the model. We can evaluate different models with varying numbers of lagged terms, similar to when we used the adjusted R-squared to assess different models with varying explanatory variables, as we saw in Chapter 6.

If we know the model's variance $\hat{\sigma}^2 = \sqrt{RSS/(n-k-1)}$, as seen in (5.12), we can compute several information criteria. These criteria are often provided in the output of regression analysis in software like Gretl, and they can be easily obtained from a regression in R. Below are the formulas for Akaikes's information criterion (Akaike, 1974), Schwarz's Bayesian information criterion (Schwarz, 1978), and the Hannan-Quinn criterion (Hannan & Quinn, 1979):

$$\text{AIC} = \ln(\hat{\sigma}^2) + \frac{2(k+1)}{T} \tag{9.5}$$

$$\text{BIC} = \ln(\hat{\sigma}^2) + \frac{k+1}{T} \ln T \tag{9.6}$$

$$\text{HQIC} = \ln(\hat{\sigma}^2) + \frac{2(k+1)}{T} \ln(\ln T) \tag{9.7}$$

Here, k represents the number of coefficients in the model, and T is the sample size. Smaller values of these information criteria indicate a better model for predicting the dependent variable (i.e. opposite R-squared which we would like to have as high as possible).

9.2.3 *AR Models in Gretl and R*

To add lagged terms to the dependent variable as explanatory variables in a model, you can click on `lags...` when specifying the model and then select the number of lagged terms for the explanatory variable (as well as for other explanatory variables if necessary), as illustrated in Fig. 9.10.

We then get the estimates similar to those shown for the autoregressive model of order one in Sect. 9.2.1:

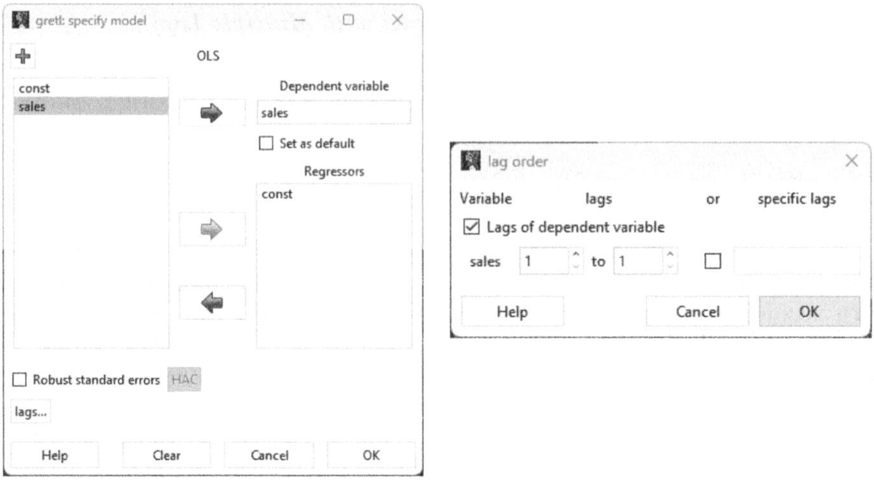

Fig. 9.10 Lagged terms in Gretl

Model 8: OLS, using observations 2000:2–2002:4 ($T = 11$)
Dependent variable: sales

	Coefficient	*Std. error*	*t-ratio*	*p-value*
Const	5,347.68	2,852.09	1.875	0.0935*
sales_1	0.721170	0.190409	3.787	0.0043***

Mean dependent var	15,181.82	S.D. dependent var	5,980.271
Sum squared resid	1.38e+08	S.E. of regression	3,914.021
R^2	0.614480	Adjusted R^2	0.571644
$F(1, 9)$	14.34507	P-value (F)	0.004299
Log-likelihood	−105.5002	Akaike criterion	215.0003
Schwarz criterion	215.7961	Hannan–Quinn	214.4987
$\hat{\rho}$	−0.285055	Durbin's h	−1.219322

9.2.4 AR Models in Gretl and R

You can also estimate an AR model by going to `Model > Univariate time series > ARIMA`. An **ARIMA model** includes both autoregressive terms, as we have discussed (AR), lagged error terms (MA, which stands for moving average), and the ability to choose the order of integration (I) for the model. Integration signifies how many times a non-stationary variable must be differenced to become stationary (we will look further into this in Sect. 9.7).

However, it's worth noting that an ARIMA model in Gretl is estimated using **maximum likelihood**, which we will not go further into in this book.

In R, you can incorporate lagged terms in a model using the `lag()` command. An AR(1) model for sales can be estimated in R with the following code:

R code 9.7: AR(1) model

```
ar1 <- lm(sales ~ lag(sales))
summary(ar1)
```

The estimated AR(1) model is thus

$$\widehat{\text{sales}}_t = 5347.59 + 0.72 \cdot \text{sales}_{t-1}$$

9.2.5 *Stationarity Test*

To test whether a variable is stationary or not, we can use a **Dickey-Fuller test**.. If we have an AR(1) model for Y_t as follows:

$$Y_t = \beta_0 + \beta_1 Y_{t-1} + u_t$$

Y_t is non-stationary if $\beta_1 \geq 1$ or $\beta_1 \leq -1$. However, this cannot be formally tested using the t-distribution because, under the null hypothesis $H_0 : \beta_1 = 1$, we have non-stationarity. Therefore, we need to use a different distribution. To make it possible to test for both a coefficient equal to -1 and 1, we need to rewrite the AR(1) model slightly. By subtracting Y_{t-1} from both sides of the equation in the AR(1) model, we get:

$$\begin{aligned}
Y_t &= \beta_0 + \beta_1 Y_{t-1} + u_t \\
Y_t - Y_{t-1} &= \beta_0 + \beta_1 Y_{t-1} - Y_{t-1} + u_t \\
\Delta Y_t &= \beta_0 + (\beta_1 - 1) Y_{t-1} + u_t \\
\Delta Y_t &= \beta_0 + \delta Y_{t-1} + u_t \qquad (9.8)
\end{aligned}$$

where $\Delta Y_t = Y_t - Y_{t-1}$ and $\delta = \beta_1 - 1$. We now want to test the hypotheses:

$$H_0 : \delta = 0\,(Y_t \text{ is non-stationary})$$
$$H_A : \delta \neq 0\,(Y_t \text{ is stationary})$$

If $\delta = 0$, we observe that $\Delta Y_t = \beta_0 + u_t$, which means that Y_t grows by the value of β_0 each period. However, with a different value for δ, the change in Y_t is influenced by the value of Y_t in the previous period. Additionally, since $\delta = \beta_1 - 1$, $\delta = 0$ implies that $\beta_1 = 1$, representing a non-stationary relationship between Y_t and Y_{t-1}. If we have negative growth for a non-stationary variable with $\beta_1 = -1$, this will be captured in the estimation of (9.8), with $\beta_0 < 0$.

However, it is also common to extend (9.8) with lagged variables of ΔY_t, resulting in:

$$\Delta Y_t = \beta_0 + \delta Y_{t-1} + \gamma_1 \Delta Y_{t-1} + \gamma_2 \Delta Y_{t-2} + \cdots + \gamma_p \Delta Y_{t-p} + u_t \qquad (9.9)$$

This model is used for conducting an **Augmented Dickey-Fuller (ADF)** test. The number of lagged variables (p) can be determined based on the type of data or using information criteria.

We can also include a trend in the model, allowing us to test whether the variable Y_t is **trend-stationary**. This means that the variable is stationary around a deterministic trend, such as a trend variable trend $= t$.

$$\Delta Y_t = \beta_0 + \delta Y_{t-1} + \gamma_1 \Delta Y_{t-1} + \gamma_2 \Delta Y_{t-2} + \cdots + \gamma_p + \gamma_t t + \Delta Y_{t-p} + u_t \qquad (9.10)$$

ADF Test in Gretl

We can conduct an ADF test in Gretl by going to `Variable > Unit root tests > Augmented Dickey-Fuller test`. Here, we can choose the number of lagged differences (which can be determined automatically using an information criterion) and decide whether we want to include a constant term (β_0) and a trend (t) in the model (Fig. 9.11).

If we set the number of lagged differences to 1, we obtain

```
Augmented Dickey-Fuller test for sales
testing down from 1 lags, criterion AIC
sample size 11
unit-root null hypothesis: a = 1

test with constant
including 0 lags of (1-L)sales
model: (1-L)y = b0 + (a-1)*y(-1) + e
estimated value of (a - 1): -0.27883
test statistic: tau_c(1) = -1.46438
asymptotic p-value 0.5519
1st-order autocorrelation coeff. for e: -0.285

with constant and trend
including one lag of (1-L)sales
model: (1-L)y = b0 + b1*t + (a-1)*y(-1) + ... + e
estimated value of (a - 1): -2.16261
test statistic: tau_ct(1) = -4.01172
asymptotic p-value 0.008385
1st-order autocorrelation coeff. for e: -0.431
```

The null hypothesis suggests non-stationarity, while the alternative hypothesis represents stationarity.

We get one test including a constant term

$$\Delta Y_t = \beta_0 + \delta Y_{t-1} + \gamma_1 \Delta Y_{t-1} + u_t$$

and one including a constant term and a trend

$$\Delta Y_t = \beta_0 + \delta Y_{t-1} + \gamma_1 \Delta Y_{t-1} + \gamma_t t + u_t$$

We observe that we keep the null hypothesis when we have only a constant term in the model, but if we include a trend, we reject the null hypothesis. This indicates that the variable is non-stationary (as indicated by the graph in Fig. 9.9), and that it is trend-stationary. If we incorporate a deterministic trend (e.g., a trend variable) into the model, we can often use regression analysis even when the variable is non-stationary.

ADF Test in R

In R, you can perform an ADF test using the `ur.df()` function from the `urca` package. You can specify the type of test you want: use `drift` if you

Fig. 9.11 Augmented Dickey-Fuller test in Gretl

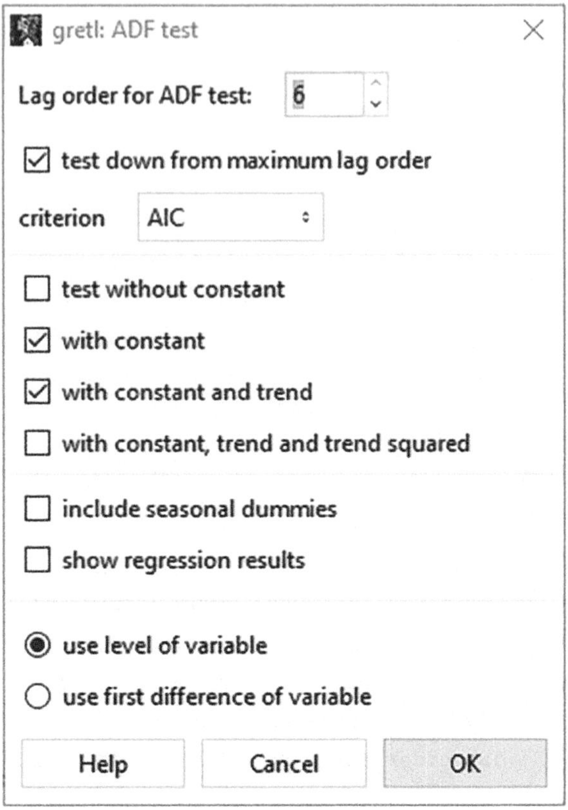

only want a constant term and `trend` if you want to include a trend in the model as well. Additionally, you can choose the number of lagged differences with the `lags=` option or use automatic lag selection based on information criteria, for example, `selectlags = c("AIC")`. For more information, refer to `help(ur.df)`.

R code 9.8: ADF test in R

```
install.packages("urca") #only necessary the first time
library(urca)
adfc <- ur.df(sales, type="drift", lags = 1)
summary(adfc)
adft <- ur.df(sales, type="trend", lags = 1)
summary(adft)
```

We then get

```
Test regression drift

Call:
lm(formula = z.diff ~ z.lag.1 + 1 + z.diff.lag)

Residuals:
Min       1Q  Median       3Q      Max
-4885.3 -2890.9  -742.7  2189.0  5380.7

Coefficients:
Estimate Std. Error t value Pr(>|t|)
(Intercept) 6009.5456  3672.6046   1.636    0.146
z.lag.1       -0.2612     0.2533  -1.031    0.337
z.diff.lag    -0.3317     0.3875  -0.856    0.420

Residual standard error: 4110 on 7 degrees of freedom
Multiple R-squared:  0.3064,Adjusted R-squared:  0.1082
F-statistic: 1.546 on 2 and 7 DF,  p-value: 0.278

Value of test-statistic is: -1.031 1.9879

Critical values for test statistics:
1pct  5pct 10pct
tau2 -3.75 -3.00 -2.63
phi1  7.88  5.18  4.12
```

```
Test regression trend

Call:
lm(formula = z.diff ~ z.lag.1 + 1 + tt + z.diff.lag)

Residuals:
Min        1Q  Median      3Q     Max
-3051.5 -1143.0  -334.1  1631.9  3232.5

Coefficients:
Estimate Std. Error t value Pr(>|t|)
(Intercept) 8606.9900  2310.4628   3.725  0.00979 **
z.lag.1        -2.1626     0.5391  -4.012  0.00702 **
tt          3554.3458   966.9468   3.676  0.01038 *
z.diff.lag      0.6494     0.3537   1.836  0.11602
---
Signif. codes:  0 '***' 0.001 '**' 0.01 '*' 0.05
                '.' 0.1 ' ' 1
Residual standard error: 2462 on 6 degrees of freedom
Multiple R-squared:  0.7867,Adjusted R-squared:  0.6801
F-statistic: 7.377 on 3 and 6 DF,  p-value: 0.01945

Value of test-statistic is: -4.0117 8.198 8.2374

Critical values for test statistics:
1pct  5pct 10pct
tau3 -4.38 -3.60 -3.24
phi2  8.21  5.68  4.67
phi3 10.61  7.24  5.91
```

We do not receive the p-value of the ADF test in the R output, but we must compare the test statistic using the first one, `tau`, which is the test statistic for δ in the ADF regression, with the critical value provided. We thus keep the null hypothesis of non-stationarity when using `type=drift`, while we reject the null hypothesis when `type=trend`, indicating that the variable is trend-stationary.

9.3 Seasonal Variation

The result of the ADF test for the variable $sales_t$ with quarterly data for $T = 12$ periods indicated that the variable had a trend-stationary behavior. By estimating an AR(1) model with a trend (often referred to as an AR(1) model with **drift**) for quarterly sales, we get:

$$sales_t = \beta_0 + \beta_1 sales_{t-1} + \gamma t + u_t \tag{9.11}$$

we obtain the following results:

Model 2: OLS, using observations 2000:2–2002:4 ($T = 11$)
Dependent variable: sales

	Coefficient	*Std. error*	*t-ratio*	*p-value*
const	4,110.77	2,015.78	2.039	0.0758*
time	2,267.00	693.836	3.267	0.0114**
sales_1	−0.351852	0.354011	−0.9939	0.3494

Mean dependent var	15,181.82	S.D. dependent var	5,980.271
Sum squared resid	59,061,616	S.E. of regression	2,717.113
R^2	0.834856	Adjusted R^2	0.793570
$F(2, 8)$	20.22124	P-value (F)	0.000744
Log-likelihood	−100.8374	Akaike criterion	207.6748
Schwarz criterion	208.8685	Hannan–Quinn	206.9224
$\hat{\rho}$	−0.219944	Durbin's h	NA

In the table below, we observe that the residuals are particularly high (and positive) in the third quarter of 2000, 2001, and 2002. This could be indicative of seasonal effects in sales, such as higher sales in the third quarter compared to the rest of the year. Products that are more popular during the summer compared to the rest of the year (for example beach supplies) might show such a pattern.

	sales	fitted	residual
2000:2	6000.00	7237.37	−1237.37
2000:3	12000.00	8800.67	3199.33
2000:4	8000.00	8956.57	−956.57
2001:1	10000.00	12630.98	−2630.98
2001:2	13000.00	14194.28	−1194.28
2001:3	20000.00	15405.72	4594.28
2001:4	15000.00	15209.76	−209.76
2002:1	17000.00	19236.03	−2236.03
2002:2	20000.00	20799.33	−799.33
2002:3	25000.00	22010.77	2989.23
2002:4	21000.00	22518.52	−1518.52

Forecast evaluation statistics (T = 11)

Mean Error	1.1575e-012
Mean Squared Error	2317.2
Root Mean Squared Error	48.137
Mean Absolute Error	1960.5
Mean Percentage Error	−2.9333
Mean Absolute Percentage Error	14.131
Theil's U_2	0.57943

9.3.1 *Seasonal Variation in Gretl and R*

To account for seasonal variation, you can create dummy variables for the different periods. In Gretl, you can do this by selecting `Add > Periodic dummies` (do not choose to center the periodic dummies as this will not be covered here). For quarterly data, this would generate variables like `dq1`, `dq2`, `dq3`, and `dq4` for the first through fourth quarters (Fig. 9.12). If you were working with monthly data, you would have 12 dummy variables.

In R, you can use the `seasonaldummy()` function from the `forecast` package to create seasonal dummy variables for a time series variable. This function will generate a matrix of dummy variables for all seasons except one (to avoid perfect multicollinearity—see Chapter 7).

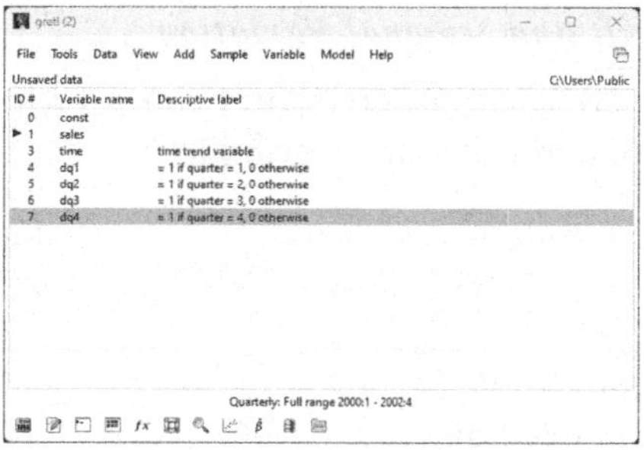
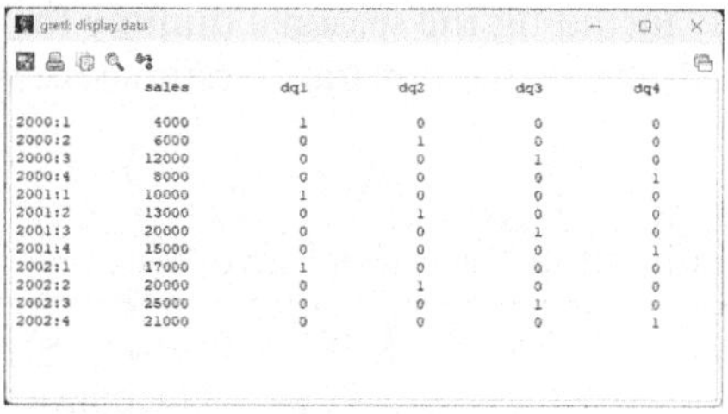

Fig. 9.12 Dummy variables for season, variable list, and dataset

R code 9.9: Seasonal variables in R

```
sales <- c(4000,6000,12000,8000,10000,13000,20000,15000,
           17000,20000,25000,21000)
sales <- ts(sales, frequency = 4, start=c(2000,1))

install.packages("forecast") #only necessary the first time
library(forecast)
season <- seasonaldummy(sales)
```

This gives

```
> sesong
      Q1 Q2 Q3
 [1,]  1  0  0
 [2,]  0  1  0
 [3,]  0  0  1
 [4,]  0  0  0
 [5,]  1  0  0
 [6,]  0  1  0
 [7,]  0  0  1
 [8,]  0  0  0
 [9,]  1  0  0
[10,]  0  1  0
[11,]  0  0  1
[12,]  0  0  0
```

9.3.2 AR(1) Model with Drift and Seasonal Variation

By including the seasonal dummy for the 3rd quarter in the AR(1) model with a drift that we previously estimated, we obtain the following model:

$$\text{sales}_t = \beta_0 + \beta_1 \text{sales}_{t-1} + \beta_2 dq3_t + \gamma t + u_t \quad (9.12)$$

Estimating this model gives us:

Model 3: OLS, using observations 2000:2–2002:4 ($T = 11$)
Dependent variable: sales

	Coefficient	*Std. error*	*t-ratio*	*p-value*
const	2,573.73	597.879	4.305	0.0035***
time	1,967.30	200.655	9.804	0.0000***
dq3	5,086.36	533.030	9.542	0.0000***
sales_1	−0.187016	0.102582	−1.823	0.1111

Mean dependent var	15,181.82	S.D. dependent var	5,980.271
Sum squared resid	4,216,260	S.E. of regression	776.0946
R^2	0.988211	Adjusted R^2	0.983158
$F(3, 7)$	195.5873	P-value (F)	4.12e–07
Log-likelihood	−86.31942	Akaike criterion	180.6388
Schwarz criterion	182.2304	Hannan–Quinn	179.6356
$\hat{\rho}$	−0.063136	Durbin's h	−0.222684

The estimated model indicates that the expected change in sales for a quarter compared to the previous quarter is 1,967.30. Thus, we can expect sales to increase by 1,967.30 units every quarter, starting from the second quarter of 2000, where the expected sales are 2,573.73 (the estimated intercept). We also find a small, non-significant effect of the autoregressive term at −0.187. The dummy variable for the third quarter has an estimated coefficient of 5,086.36, which means that, compared to the other quarters and the other explanatory variables in our model (previous period's sales and the linear trend), we expect sales to be 5,086 units higher in the third quarter.

9.4 Forecasts and Model Evaluation

Just like in Sects. 5.4 and 5.6.4, we can create forecasts for the expected value of the dependent variable and a prediction interval for a time series variable. For the quarterly sales data, we estimated three different models: (1) an AR(1) model, (2) an AR(1) model with a drift, and (3) an AR(1) model with a drift and a dummy variable for the first quarter. We had the estimated models as follows:

$$\widehat{\text{sales}_t} = 5347.7 + 0.72\text{sales}_{t-1}$$

$$\widehat{\text{sales}}_t = 4110.8 - 0.35\text{sales}_{t-1} + 2267.0 \cdot t$$

$$\widehat{\text{sales}}_t = 2573.7 - 0.19\text{sales}_{t-1} + 1967.3 \cdot t + 5086.3 \cdot dq3_t$$

We do not have observations for the first quarter of 2003, but we can create forecasts for the first quarter of 2001 using each of the three models. Here, we have $t = 13$, $\text{sales}_{t-1} = \text{sales}_{12} = 21000$, and $dq3_{13} = 0$. We then get:

$$\widehat{\text{sales}}_{13} = 5347.7 + 0.72\text{sales}_{12}$$

$$\widehat{\text{sales}}_{13} = 4110.8 - 0.35\text{sales}_{12} + 2267.0 \cdot 13$$

$$\widehat{\text{sales}}_{13} = 2573.7 - 0.19\text{sales}_{12} + 1967.3 \cdot 13 + 5086.3 \cdot 0$$

which results in:

$$\widehat{\text{sales}}_{13} = 5347.7 + 0.72 \cdot 21000 = 20,468$$

$$\widehat{\text{sales}}_{13} = 4110.8 - 0.35 \cdot 21000 + 2267.0 \cdot 13 = 26232$$

$$\widehat{\text{sales}}_{13} = 2573.7 - 0.19 \cdot 21000 + 1967.3 \cdot 13 + 0 = 24159$$

Here, we see that the three models have different predictions for sales in the first quarter of 2003. Since we do not know the sales in the first quarter of 2003, we cannot assess which model provides the best forecast. However, we can evaluate a model based on the quality of its forecasts for periods already included in the dataset. Then we can examine how the model's predictions compare to observed values. Since all three models use a lagged observation, we can make predictions from the second quarter of 2000 to the fourth quarter of 2002, which we can compare with observed values. For the AR(1) model, we get, for example, a predicted value for the second quarter of 2000:

$$\widehat{\text{sales}}_2 = 5347.7 + 0.72\text{sales}_1 = 5347.7 + 0.72 \cdot 6000 = 8227.7$$

We summarize these in Table 9.7.

The residuals for each model are shown in Table 9.8

The residuals for each period are the forecast errors and can be used to evaluate how well each model predicts sales. The residuals $\hat{u}_t$ measure the **forecast error**. We can sum these deviations in different ways to obtain various metrics of forecast error commonly used in model evaluations. One such metric is the root mean squared error (RMSE), which involves squaring all deviations and calculating the square root of the mean squared deviation:

$$\text{RMSE} = \sqrt{\frac{1}{TT} \sum \hat{u}_t^2} \tag{9.13}$$

where TT is the number of predictions made (in our example, 11). Additionally, we can calculate the absolute value of each deviation and the Mean

Table 9.7 Actual sales and predicted sales with AR(1) model (1), AR(1) model with drift (2), and AR(1) model with drift and dummy variable for the third quarter (3)

t	*Sales*	*(1)*	*(2)*	*(3)*
1	4,000			
2	6,000	8,232.36	7,237.37	5,760.27
3	12,000	9,674.70	8,800.67	12,439.90
4	8,000	14,001.72	8,956.57	8,198.75
5	10,000	11,117.04	12,630.98	10,914.12
6	13,000	12,559.38	14,194.28	12,507.39
7	20,000	14,722.89	15,405.72	19,000.00
8	15,000	19,771.08	15,209.76	14,571.83
9	17,000	16,165.23	19,236.03	17,474.21
10	20,000	17,607.57	20,799.33	19,067.48
11	25,000	19,771.08	22,010.77	25,560.10
12	21,000	23,376.94	22,518.52	21,505.96

Table 9.8 Residuals for AR(1) model (1), AR(1) model with drift (2) and AR(1) model with drift and dummy variable for the third quarter (3)

t	*(1)*	*(2)*	*(3)*
2	−2,232.36	−1,237.37	239.73
3	2,325.3	3,199.33	−439.90
4	−6,001.72	−956.57	−198.75
5	−1,117.04	−2,630.98	−914.12
6	440.62	−1,194.28	492.61
7	5,277.11	4,594.28	1,000.00
8	−4,771.08	−209.76	428.17
9	834.77	−2,236.03	−474.21
10	2,392.43	−799.33	932.52
11	5,228.92	2,989.23	−560.10
12	−2,376.94	−1,518.52	−505.96

Absolute Deviation (MAD):

$$\text{MAE} = \frac{1}{TT} \sum |\hat{u}_t| \tag{9.14}$$

We can also measure the percentage deviation from the observed value Y_t to calculate the Mean Percentage Error (MPE):

$$\text{MPE} = \frac{1}{TT} \sum \frac{\hat{u}_t}{Y_t} \cdot 100 \tag{9.15}$$

Table 9.9 Calculations for forecast errors of the AR(1) model

t	*Sales*	$\hat{u}$	$\hat{u}^2$	$\lvert\hat{u}\rvert$	$\hat{u}/sales$	$\lvert\hat{u}/sales\rvert$
2	6,000	−2,232.4	4,983,431.2	2,232.4	−37.2	37.2
3	12,000	2,325.3	5,407,020.1	2,325.3	19.4	19.4
4	8,000	−6,001.7	36,020,643.0	6,001.7	−75.0	75.0
5	10,000	−1,117.0	1,247,778.4	1,117.0	−11.2	11.2
6	13,000	440.6	194,146.0	440.6	3.4	3.4
7	20,000	5,277.1	27,847,890.0	5,277.1	26.4	26.4
8	15,000	−4,771.1	22,763,204.4	4,771.1	−31.8	31.8
9	17,000	834.8	696,841.0	834.8	4.9	4.9
10	20,000	2,392.4	5,723,721.3	2,392.4	12.0	12.0
11	25,000	5,228.9	27,341,604.4	5,228.9	20.9	20.9
12	21,000	−2,376.9	5,649,843.8	2,376.9	−11.3	11.3
Sum		0.0	137,876,123.3	32,998.3	−79.6	253.5
Mean			12,534,193.0	2,999.8	−7.2	23.0

Furthermore, we can measure the absolute percentage deviation from the observed value to calculate the Mean Absolute Percentage Error (MAPE):

$$\text{MAPE} = \frac{1}{TT}\sum\left|\frac{\hat{u}_t}{Y_t}\right| \cdot 100 \tag{9.16}$$

Smaller values for these metrics indicate a better model. We can calculate these metrics by finding absolute deviations, squared deviations, percentage deviations, and the absolute value of percentage deviations. This is shown for the AR(1) model in Table 9.9.

From the table, we see that

$$\begin{aligned} \text{RMSE} &= \sqrt{12534193.0} = 3540.4 \\ \text{MAE} &= 2999.8 \\ \text{MPE} &= -7.2 \\ \text{MAPE} &= 23.0 \end{aligned}$$

For the AR(1) model with drift, we get

$$\begin{aligned} \text{RMSE} &= 2317.2 \\ \text{MAE} &= 1960.5 \\ \text{MPE} &= -2.9 \\ \text{MAPE} &= 14.0 \end{aligned}$$

and for the AR(1) model with drift and a dummy variable for the third quarter, we get

$$\text{RMSE} = 619.11$$

$$\text{MAE} = 562.37$$
$$\text{MPE} = -0.22$$
$$\text{MAPE} = 3.91$$

All of the metrics indicate that the AR(1) model with a drift and a dummy variable for the third quarter is best suited to predict sales.

9.4.1 *Evaluating Forecasts in Gretl and R*

If we go to `Analysis > Display actual, fitted, residual` in Gretl, we get observed values, predicted values and residuals, in addition to values for RMSE, MAD, MPE and MAPE. For the AR(1) model, we get

Model estimation range: 2000:2–2002:4
Standard error of residuals = 3914.02

	sales	fitted	residual
2000:2	6000.00	8232.36	−2232.36
2000:3	12000.00	9674.70	2325.30
2000:4	8000.00	14001.72	−6001.72
2001:1	10000.00	11117.04	−1117.04
2001:2	13000.00	12559.38	440.62
2001:3	20000.00	14722.89	5277.11
2001:4	15000.00	19771.08	−4771.08
2002:1	17000.00	16165.23	834.77
2002:2	20000.00	17607.57	2392.43
2002:3	25000.00	19771.08	5228.92
2002:4	21000.00	23376.94	−2376.94

Forecast evaluation statistics (T = 11)

Mean Error	8.2681e-013
Mean Squared Error	3540.4
Root Mean Squared Error	59.501
Mean Absolute Error	2999.8
Mean Percentage Error	−7.2348
Mean Absolute Percentage Error	23.042
Theil's U_2	0.65135

These numbers are in line with what we calculated earlier.

In R, we can use the `MLmetrics` package and the functions `RMSE()`, `MAE()`, and `MAPE()` to calculate RMSE, MAE, and MAPE, respectively. We input the vector of predicted values and the vector of actual values inside the parentheses.

R code 9.10: Evaluating forecasts for the AR(1) model in R

```
install.packages("MLmetrics") #only the first time
library(MLmetrics)
ar1 <- lm(sales ~ lag(sales))
ar1tr <- lm(sales ~ lag(sales) + trend12)
ar1dseas <- lm(sales ~ lag(sales) + trend12 + dq3)
RMSE(ar1$fitted.values, sales[2:12])
MAE(ar1$fitted.values, sales[2:12])
MAPE(ar1$fitted.values, sales[2:12])
```

This provides

```
> RMSE(ar1$fitted.values, sales[2:12])
[1] 3540.366
> MAE(ar1$fitted.values, sales[2:12])
[1] 2999.844
> MAPE(ar1$fitted.values, sales[2:12])
[1] 0.2304223
```

Note that MAPE is given in rates and not in percentages in R, in contrast to what we got when we calculated manually and the output from Gretl.

9.5 Testing for Autocorrelation in the Residuals

To meet the assumptions of the OLS model, the error term must be both stationary and non-autocorrelated. This means that the correlation between u_t and the errors from previous periods must be zero, i.e., $\text{Corr}(u_t, u_{t-1}) = 0$, $\text{Corr}(u_t, u_{t-2}) = 0$, and so on. If it turns out that we have autocorrelated errors, this can often be addressed by introducing more dynamics into the model, such as including additional lagged values of the dependent variable or other explanatory variables.

We can estimate the model

$$\hat{u}_t = \beta_0 + \gamma_1 Y_{t-1} + \cdots + \gamma_p Y_{t-p} + \beta_1 \hat{u}_{t-1} + \beta_2 \hat{u}_{t-2} + \cdots + \beta_3 \hat{u}_{t-q} + \varepsilon_t \quad (9.17)$$

where $\hat{u}_t$ is the error term from a model we have estimated, q is the number of lagged periods (which can be determined, for example, by information criteria or based on the dataset's frequency), p is the number of lagged periods in the model we have estimated, and ε_t is the error term of this model.

We can formulate the hypotheses as follows:

$$H_0 : \gamma_1 = \cdots = \gamma_p = \beta_1 = \cdots = \beta_q = 0 \tag{9.18}$$

$$H_A : \gamma_1 \neq 0, \ldots \text{ and/or } \gamma_p \neq 0, \text{ and/or } \beta_1 \neq 0, \text{ and/or } \ldots \beta_q \neq 0 \tag{9.19}$$

H_0 then implies that the error term is not autocorrelated, while H_A implies that there is autocorrelation in the error term. This F-test for autocorrelated errors is known as the **Breusch–Godfrey test**. The test statistic is given by:

$$F = \frac{(R^2 - R^2_{\text{null}})/m}{(1 - R^2)/(n - p - q - 1)} \tag{9.20}$$

where, under the null hypothesis, the model will be $\hat{u}_t = \beta_0 + \varepsilon_t$, resulting in $R^2_{\text{null}} = 0$. Here, m is the number of equal signs in the null hypothesis, q is the number of lagged periods for the residuals, and p is the number of lagged periods in the estimated model.

9.5.1 Testing for Autocorrelated Residuals in Gretl and R

In Gretl, we can conduct the Breusch–Godfrey test by going to `Tests > Autocorrelation`. We need to select the number of lags, which we choose to be $q = 4$ here since we have quarterly data. For the AR(1) model, the results are as follows:

```
Breusch-Godfrey test for autocorrelation up to order 4
OLS, using observations 2000:2-2002:4 (T = 11)
Dependent variable: uhat

coefficient    std. error    t-ratio   p-value
----------------------------------------------------------
const        -4212.47        3502.64      -1.203    0.2829
sales_1          0.329459       0.239538    1.375    0.2274
uhat_1        -0.578697      0.468005   -1.237    0.2712
uhat_2        -0.377319      0.446705   -0.8447   0.4368
uhat_3        -0.241450      0.456505   -0.5289   0.6195
uhat_4         0.665984      0.436741    1.525    0.1878

Unadjusted R-squared = 0.626431

Test statistic: LMF = 2.096099.
with p-value = P(F(4.5) > 2.0961) = 0.219

Alternative statistic: TR^2 = 6.890738.
with p-value = P(Chi-square(4) > 6.89074) = 0.142

Ljung-Box Q' = 9.09834.
with p-value = P(Chi-square(4) > 9.09834) = 0.0587
```

We observe that we have a p-value of 0.219 according to the F-test, which means we keep the null hypothesis of no autocorrelation in the errors. Note that the Breusch–Godfrey test conducted in Gretl also includes a lag of the dependent variable.

In R, we can use the command `bgtest(model, order = q, type = "F")` in the `lmtest` package, where `model` is the name of the model we want to test, q is the number of lags we use, and `type = "F"` indicates that we are using an F-test (the default setting is a chi-squared test).

R code 9.11: Breusch-Godfrey-test of the AR(1) model in R

```
install.packages("lmtest") #only necessary the first time
library(lmtest)
bgtest(ar1,order=4,type="F")
```

This gives

```
Breusch-Godfrey test for serial correlation
of order up to 4


data:  ar1
LM test = 2.0961, df1 = 4, df2 = 5, p-value = 0.2189
```

9.6 Parameter Stability

For time series data, it is important to make sure that the estimated parameters are stable over time. If there are certain events outside of the regression model that has changed the relationship between the variables in the model, this may affect **parameter stability**.

Let us now assume that we have extended the time series data for sales and prices in the example we used previously in the chapter. The data now goes from 2001 to 2012 and covers 12 years ($T = 12$), as shown in Fig. 9.13.

While prices seem to increase steadily over the entire sample, sales seem to increase until 2006, while it is relatively steady from 2007 to 2012. The relationship between sales and prices may thus not be the same across the sample. If we estimate the model

$$\text{sales}_t = \beta_0 + \beta_1 \text{price}_t + u_t$$

for the entire sample, we may have parameter instability.

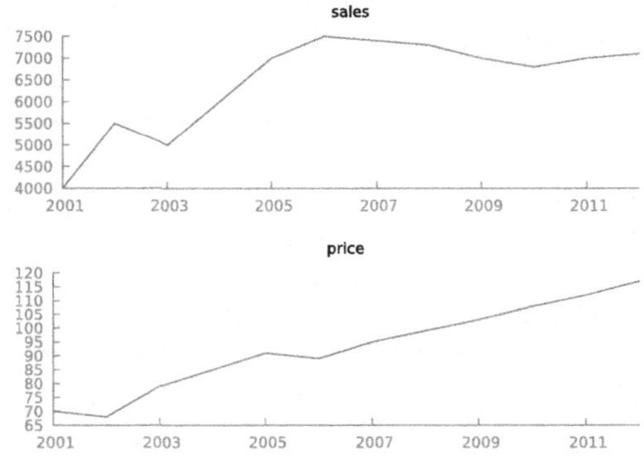

Year	t	sales	price
2001	1	4,000	70
2002	2	5,500	68
2003	3	5,000	79
2004	4	6,000	85
2005	5	7,000	91
2006	6	7,500	89
2007	7	7,400	95
2008	8	7,300	99
2009	9	7,000	103
2010	10	6,800	108
2011	11	7,000	112
2012	12	7,100	117

Fig. 9.13 Sales and price 2001–2012

This may be taken into account using dummy variables (in this case, a **shift dummy**), analogous to what we did in Chapter 7. Hence, we can estimate

$$\text{sales}_t = \beta_0 + \beta_1 \text{price}_t + \beta_2 \text{dum}_t + \beta_3 \text{dum}_t \cdot \text{price}_t + u_t$$

where dum_t is 0 for 2001–2006 and 1 for 2007–2012. We can then test

$$H_0 : \beta_2 = 0 \text{ and } \beta_3 = 0$$
$$H_A : \beta_2 \neq 0 \text{ and/or } \beta_3 \neq 0$$

using an F-test to see if we have parameter instability. This is known as the **Chow test**.

9.6.1 *Chow Test in Gretl*

After performing an OLS time series regression, we can go to `Tests` > `Chow test`, and select the period where we want to split the sample (the first period of the second part of the sample, which in our example is 2007) to conduct the test based on our suspicions. This will provide results from the estimated regression with dummy variables, and results from the Chow test (which is the F-test on the hypothesis stated above).

```
Augmented regression for Chow test
OLS, using observations 2001-2012 (T = 12)
Dependent variable: sales

coefficient   std. error   t-ratio   p-value
---------------------------------------------------------
const        -3001.06       2194.79      -1.367   0.2087
price          109.972        27.1562     4.050   0.0037  ***
splitdum     11844.9        4038.83       2.933   0.0189  **
sd_price      -126.475        41.9737    -3.013   0.0167  **

Mean dependent var  6466.667   S.D. dependent var   1103.163
Sum squared resid    2780716   S.E. of regression   589.5673
R-squared           0.792277   Adjusted R-squared   0.714381
F(3, 8)             10.17095   P-value(F)           0.004183
Log-likelihood     -91.14714   Akaike criterion     190.2943
Schwarz criterion   192.2339   Hannan-Quinn         189.5762
rho                -0.384544   Durbin-Watson        2.572148

Chow test for structural break at observation 2007
F(2, 8) = 4.58022 with p-value 0.0472
```

The result shows a test statistic of $F = 4.58$ and a p-value just below 0.05, such that we conclude that we reject the null hypothesis of parameter stability at a 5% significance level, suggesting that we have parameter instability. In fact, the estimated effect of price on sales after 2006 is negative: $\hat{\beta}_1 + \hat{\beta}_3 = 109.97 - 126.48 = -16.51$.

9.6.2 *Chow Test in R*

In R, we can use the command `sctest` to conduct a Chow test. This is found in the `strucchange` library.

R code 9.12: Chow test

```
sales2 <- c(4000,5500,5000,6000,7000,7500,7400,7300,7000,
6800,7000,7100)
price2 <- c(70,68,79,85,91,89,95,99,103,108,112,117)

#install.packages("strucchange")
library(strucchange)
sctest(sales2 ~ price2, type = "Chow", point = 6)
```

As seen in the code, we need to insert `point` = 6 as the period when the first part of the sample ends (2006 or $t = 6$).

```
Chow test


data:  sales2 ~ price2
F = 4.5802, p-value = 0.04723
```

The result shows a test statistic of $F = 4.58$ and a p-value just below 0.05, such that we conclude that we reject the null hypothesis of parameter stability at a 5% significance level, suggesting that we have parameter instability.

9.7 Cointegration

As we saw previously in this chapter when estimating the regression model with quantity sold and price,

$$\text{sales}_t = \beta_0 + \beta_1 \text{price}_t + u_t$$

we observed a very high R-squared and significant effects, possibly due to spurious regression. We got

$$\widehat{\text{sales}_t} = \underset{(3020.41)}{-30001.06} + \underset{(37.37)}{109.97}\text{price}_t$$

with $R^2 = 0.68$. This is often an issue when using non-stationary variables, namely spurious regression which we addressed in Sect. 9.1.4.

We then included a **deterministic trend**, t, in the model in (9.1) and found that the effect of price on sales was no longer significant:

$$\widehat{\text{sales}_t} = \underset{(3400.11)}{7891.37} - \underset{(53.35)}{68.81}\text{price}_t + \underset{(276.87)}{991.37}t$$

Here, we got $R^2 = 0.94$, which was even higher, but the deterministic trend explained the quantity sold and not the price, which had an insignificant effect.

Another possibility to deal with this was to use the first difference of the variables, as in (9.2). This provided a regression with non-stationary variables. We then got

$$\widehat{\Delta\text{sales}_t} = \underset{(380.88)}{1018.63} - \underset{(60.07)}{83.85}\ \Delta\text{price}_t$$

where $\Delta\text{sales}_t = \text{sales}_t - \text{sales}_{t-1}$ and $\Delta\text{price}_t = \text{price}_t - \text{price}_{t-1}$.

However, we do not get any long-run information about the relationship between the dependent variable and the regressors. Adding a trend only says something about the growth in the dependent variable and using first differences only estimates the effect of changes in the regressors on the dependent variable. In order to estimate the level effect, we can use an **error correction model**. This allows modeling both the short-run and long-run relationship jointly if the variables are **cointegrated**.

Cointegration between two non-stationary variables implies that a linear combination between the two variables is stationary, and it can be interpreted as if there is a long-run relationship between the variables. This is the case if the variables have a common **stochastic trend**. For two variables (a bivariate system), this may be tested using the **Engle-Granger test for cointegration** (Engle & Granger, 1987).

The Engle-Granger test involves testing for cointegration through the hypotheses

$$H_0 : \text{There is no cointegration between the variables}$$
$$H_A : \text{There is cointegration between the variables}$$

by investigating whether a combination between two non-stationary variables is stationary. This is done in two steps:

1. Both variables are non-stationary. This may be tested through an ADF test, as described in Sect. 9.2.5.
2. A linear combination between the variables is stationary. This can be done by testing whether the residual from the regression between the two variables is stationary.

Hence, if we estimate the model

$$Y_t = \beta_0 + \beta_1 X_t + u_t$$

we have that

$$\hat{u}_t = Y_t - \hat{\beta}_0 - \hat{\beta}_1 X_t$$

If Y_t and X_t are non-stationary, we can test for cointegration between these two variables by testing whether $\hat{u}_t$ is stationary. This is illustrated in Gretl and R below.

9.7.1 Cointegration in Gretl

To test for cointegration between variables in Gretl using the Engle-Granger test, go to `Model > Multivariate time series > Cointegration test (Engle-Granger)`, and then select the variables that we want to test for cointegration. We can also choose the number of lags and whether or not we want to include a constant or a trend in the ADF tests.

```
Step 1: testing for a unit root in sales

Augmented Dickey-Fuller test for sales
including one lag of (1-L)sales
sample size 4
unit-root null hypothesis: a = 1

test with constant
model: (1-L)y = b0 + (a-1)*y(-1) + ... + e
estimated value of (a - 1): 0.384615
test statistic: tau_c(1) = 0.652174
asymptotic p-value 0.9911
1st-order autocorrelation coeff. for e: -0.378

Step 2: testing for a unit root in price

Augmented Dickey-Fuller test for price
including one lag of (1-L)price
sample size 4
unit-root null hypothesis: a = 1

test with constant
model: (1-L)y = b0 + (a-1)*y(-1) + ... + e
estimated value of (a - 1): -0.532241
test statistic: tau_c(1) = -2.04488
asymptotic p-value 0.2677
1st-order autocorrelation coeff. for e: -0.770
```

```
Step 3: cointegrating regression

Cointegrating regression -
OLS, using observations 2001-2006 (T = 6)
Dependent variable: sales

coefficient   std. error   t-ratio   p-value
-------------------------------------------------------
const        -3001.06      3020.41      -0.9936   0.3767
price          109.972       37.3717      2.943   0.0423 **

Mean dependent var  5833.333  S.D. dependent var  1290.994
Sum squared resid    2633133  S.E. of regression  811.3466
R-squared           0.684024  Adjusted R-squared  0.605030
Log-likelihood     -47.48941  Akaike criterion    98.97882
Schwarz criterion   98.56233  Hannan-Quinn        97.31161
rho                -0.555494  Durbin-Watson       2.518289

Step 4: testing for a unit root in uhat

Augmented Dickey-Fuller test for uhat
including one lag of (1-L)uhat
sample size 4
unit-root null hypothesis: a = 1

test without constant
model: (1-L)y = (a-1)*y(-1) + ... + e
estimated value of (a - 1): -1.89346
test statistic: tau_c(2) = -1.74624
asymptotic p-value 0.6565
1st-order autocorrelation coeff. for e: 0.073
```

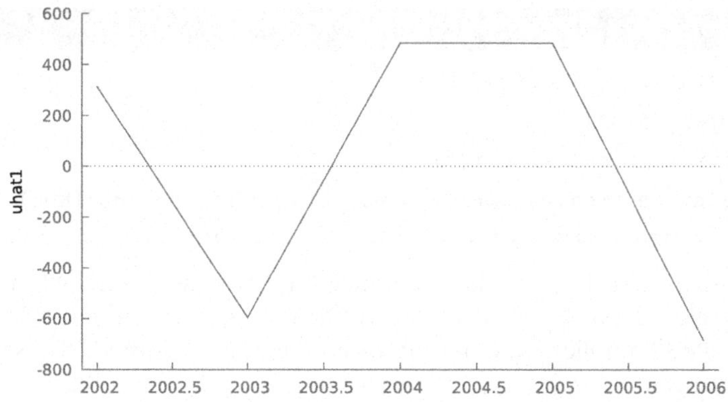

Fig. 9.14 Residuals from OLS regression on $\text{sales}_t = \beta_0 + \beta_1 \text{price}_t + u_t$

```
There is evidence for a cointegrating relationship if:
(a) The unit-root hypothesis is not rejected
for the individual variables, and
(b) the unit-root hypothesis is rejected
for the residuals (uhat) from the cointegrating regression.
```

From the Gretl output, we see that both sales and price are non-stationary, since we are not able to reject the null hypothesis of a unit root (Step 1 and Step 2 in the output). From the cointegrating regression (Step 3), we the OLS estimates of $\text{sales}_t = \beta_0 + \beta_1 \text{price}_t + u_t$ similar to what we found previously in the chapter. In Step 4, the ADF test for stationarity of the residual from the OLS regression is conducted. The result shows that the null hypothesis of non-stationarity cannot be rejected, implying that there is no cointegration between price and sales and that they do not share a common stochastic trend. As shown in Fig. 9.14, the short sample size of only $T = 5$ observations makes it hard to assess this graphically, and it also affects the critical values for the ADF test.

9.7.2 Cointegration in R

In R, we can test for cointegration using the Engle-Granger test by the function `coint.test` in the package `aTSA`.

R code 9.13: Cointegration test

```
#install.packages("aTSA")
library(aTSA)
coint.test(sales, price)
#coint.test(sales, price, d=0, nlag=NULL, output=TRUE)
```

`d=0` means zero lags for the difference (and correspondingly 1 lag for the levels), while `nlag=NULL` is the lag order to calculate the test statistic and `output=TRUE` implies that output from the test is printed. These are the default settings.

This provides

```
Response: sales
Input: price
Number of inputs: 1
Model: y ~ X + 1
-------------------------------
Engle-Granger Cointegration Test
alternative: cointegrated

Type 1: no trend
lag      EG p.value
2.00   -1.75    0.10
-----
Type 2: linear trend
lag      EG p.value
2.000  -0.111   0.100
-----
Type 3: quadratic trend
lag      EG p.value
2     NaN     NaN
-----------
Note: p.value = 0.01 means p.value <= 0.01
: p.value = 0.10 means p.value >= 0.10
```

The ADF test statistic from the ADF test on the residual from the OLS regression is shown under EG, and for no trend we get a test statistic of -1.75. The p-value is shown to be larger than or equal to 0.10 thus keeping the null of a unit root in the residual and no cointegration between sales and price.

References

Akaike, H. (1974). A new look at the statistical model identification. *IEEE Transactions on Automatic Control, 19*(6), 716–723.

Engle, R. F., & Granger, C. W. (1987). Co-integration and error correction: Representation, estimation, and testing. *Econometrica: Journal of the Econometric Society, 55*(2), 251–276.

Hannan, E. J., & Quinn, B. G. (1979). The determination of the order of an autoregression. *Journal of the Royal Statistical Society: Series B (Methodological), 41*(2), 190–195.

Johnson, S. R. (2008). The trouble with QSAR (or how I learned to stop worrying and embrace fallacy). *Journal of Chemical Information and Modeling, 48*(1), 25–26.

Schwarz, G. (1978). Estimating the dimension of a model. *Annals of Statistics, 6*(2), 461–464.

CHAPTER 10

Other Statistical Tools

Even though we can use the regression model presented in the previous chapters to investigate many interesting issues, other statistical tools are also often used. This chapter presents t-tests, chi-square tets, and analysis of variance, in addition to showing how we may perform these tests in a regression model using what we have learned in the previous chapters of the book.

We have primarily focused on the regression model to analyze relationships so far. However, there are also other statistical tools commonly used to explore relationships. For some analyses, it may be more convenient to employ these tools rather than a regression model. However, it is also possible to utilize the regression model to perform many of the tests in this chapter.

10.1 T-TEST

To examine differences between the mean values of two different variables or groups, we can use a **t-test**, often referred to as the Student's t-test or the unpaired t-test. If we have two variables or groups, denoted as A and B, and we want to investigate whether they have different means, we can formulate the hypotheses as follows:

$$H_0\colon \mu_A = \mu_B$$
$$H_A\colon \mu_A \neq \mu_B$$

Supplementary Information The online version contains supplementary material available at https://doi.org/10.1007/978-3-031-53142-2_10.

B. Karlsen Kivedal, *Applied Statistics and Econometrics*, https://doi.org/10.1007/978-3-031-53142-2_10

where μ_A represents the population mean of variable A, and μ_B represents the population mean of variable B. If we have two groups (or time periods), we often want to test for differing means. Even though there may be reasons for the population standard deviations of the two groups to be different, the test presented here assumes that they are equal.

The test statistic can be calculated as

$$T = \frac{\bar{X}_A - \bar{X}_B}{S\sqrt{1/n_A + 1/n_B}} \tag{10.1}$$

where $\bar{X}_A$ and $\bar{X}_B$ represent the sample means of variable or group A and B, and n_A and n_B are the sample sizes for the two groups. It is important to note that the groups do not need to have the same sample size to perform such a test. S is the estimated standard deviation assumed to be the same in both groups in the population,

$$S = \sqrt{\frac{(n_A - 1)S_A^2 + (n_B - 1)S_B^2}{n_A + n_B - 2}} \tag{10.2}$$

where S_A^2 and S_B^2 represent the sample variances (the squared sample standard deviations) of group A and group B, respectively.

10.1.1 Example

If we consider the individuals from our previous example, we had four people employed in the public sector and four in the private sector. We can treat these groups separately, as shown in Table 10.1.

Here, Wpub represents the wage for the four employees in the public sector, and Wpriv represents the wage for the four employees in the private sector. We can now use a t-test to determine if there is a significant difference in the mean wage between the private and public sectors based on our sample of eight individuals, with four from each sector.

Example by Hand

To calculate the test statistic for the t-test, we need to compute the standard deviation S that applies to both groups collectively. First, we need to find the standard deviation for the two variables separately. If we use A for the public

Table 10.1 Wage for employees in the public and private sectors

Public Sector Wage (Wpub)	*Private Sector Wage (Wpriv)*
320	480
290	510
170	410
370	650

Table 10.2 Wage in the public and private sectors

X_A	X_B	$(X_A - \bar{X}_A)$	$(X_A - \bar{X}_A)^2$	$(X_B - \bar{X}_B)$	$(X_B - \bar{X}_B)^2$
320	480	32.5	1,056.25	−32.5	1,056.25
290	510	2.5	6.25	−2.5	6.25
170	410	−117.5	13,806.25	−102.5	10,506.25
370	650	82.5	6,806.25	137.5	18,906.25
1 150	2,050	0	21,675	0	30,475

sector and B for the private sector, and thus the variables X_A and X_B, we have the sample means

$$\bar{X}_A = \frac{320 + 290 + 170 + 370}{4} = 287.5$$

$$\bar{X}_B = \frac{480 + 510 + 410 + 650}{4} = 512.5$$

We can then calculate various deviations, as shown in Table 10.2.

This allows us to calculate the standard deviation for each group:

$$S_A = \sqrt{\frac{\sum_{i=1}^{4}(W_{\text{pub},i} - \bar{W}_{\text{pub}})}{n-1}} = \sqrt{\frac{21675}{3}} = \sqrt{7225} = 85.00$$

$$S_B = \sqrt{\frac{\sum_{i=1}^{4}(W_{\text{priv},i} - \bar{W}_{\text{priv}})}{n-1}} = \sqrt{\frac{30475}{3}} = \sqrt{10158.33} = 100.79$$

The estimate for the overall standard deviation is then:

$$S = \sqrt{\frac{(n_A - 1)S_A^2 + (n_B - 1)S_B^2}{n_A + n_B - 2}}$$

$$= \sqrt{\frac{(4-1)\cdot 85^2 + (4-1)\cdot 100.79^2}{4+4-2}} = 93.229$$

And the test statistic is:

$$T = \frac{\bar{W}_{\text{pub}} - \bar{W}_{\text{priv}}}{S\sqrt{1/n_A + 1/n_B}}$$

$$= \frac{287.5 - 512.5}{93.229 \cdot \sqrt{1/4 + 1/4}} = -3.413$$

Critical values (in this case, a two-tailed hypothesis test) are found in the t-distribution with $n_A + n_B - 2$ degrees of freedom. For a significance level of

5

$$cv = \pm t_{\alpha/2}(n_A + n_B - 2) = \pm t_{0.025}(6) = \pm 2.447$$

The test statistic is larger in absolute value than the critical value, so we reject the null hypothesis. We conclude that there is a difference in wages between the public and private sectors.

Example in Gretl and R
In Gretl, we can perform a t-test between two variables by going to `Tools > Test statistics calculator`, where we select the `2 means` tab and choose the two variables we want to test against each other. We also need to check the `Assume common pop. std. dev` option since we assume that the standard deviation is the same in both groups here.

We then get:

```
Null hypothesis: Difference of means = 0

Sample 1:
n = 4, mean = 287.5, s.d. = 85
standard error of mean = 42.5
95% confidence interval for mean: 152.246 to 422.754

Sample 2:
n = 4, mean = 512.5, s.d. = 100.789
standard error of mean = 50.3943
95% confidence interval for mean: 352.123 to 672.877

Test statistic: t(6) = (287.5 - 512.5)/65.9229 = -3.41308
Two-tailed p-value = 0.01426
(one-tailed = 0.007132)
```

and Figure 10.1.

In R, we can perform a t-test by using the code `t.test(X, Y, var.equal = TRUE)`, where `X` and `Y` are the two variables or groups.

R code 10.1: T-test

```
t.test(Wpub,Wpriv, var.equal=TRUE)
```

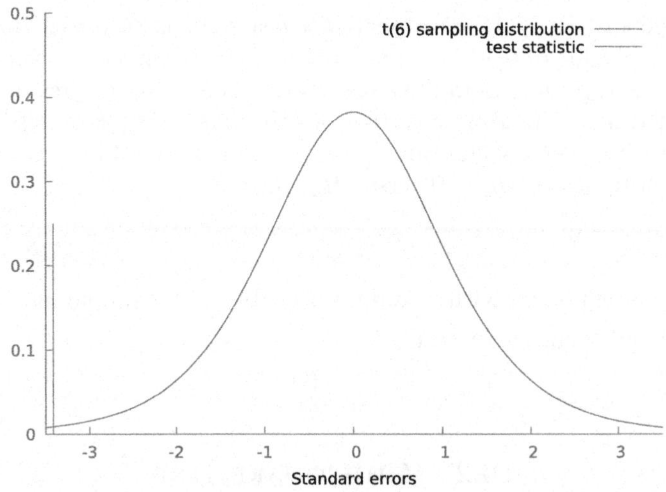

Fig. 10.1 Unpaired t-test. Figure from test in Gretl

This gives

```
Two Sample t-test

data:  Wpub and Wpriv
t = -3.4131, df = 6, p-value = 0.01426
alternative hypothesis: true difference in means
is not equal to 0
95 percent confidence interval:
-386.30761  -63.69239
sample estimates:
mean of x mean of y
287.5     512.5
```

10.1.2 T-test or Regression Model

We also tested for wage differences between the public and private sectors using regression with a dummy variable in Chapter 7. For the first two models:

$$\text{wage}_i = \beta_0 + \beta_1 \text{public}_i + \varepsilon_i$$
$$\text{wage}_i = \beta_0 + \beta_1 \text{private}_i + \varepsilon_i$$

as estimated in Sect. 7.1, we observed a test statistic of 3.413 (in absolute value) and a p-value of 0.143 for the estimated coefficients $\hat{\beta}_1$. This is the same test statistic and p-value as in the t-test above. This demonstrates that a t-test, as performed here, can also be carried out through a simple regression with a dummy variable. The conclusion remains the same, even if in a regression, we test the hypotheses H_0: $\beta_1 = 0$ versus H_A: $\beta_1 \neq 0$.

T-test or Regression

A simple regression with a dummy variable and a simple hypothesis test is equivalent to a t-test.

10.2 Chi-square Test

Chi-square tests are widely used to examine differences between various groups, especially when dealing with **categorical variables**.

We can have a dataset, or often, we only have a table showing the distribution. Let us assume that we have data for two variables for 500 individuals. One variable represents whether a person has purchased product A (with a value of 1 for yes and 0 for no), while the other variable indicates whether the person is a child (1 for child, 0 for not a child). We thus have two binary variables. This can be illustrated in Table 10.3.

To examine differences between children and adults, we calculate a table of expected values assuming the groups have the same probability distribution (Table 10.4). Testing this simultaneously is known as a Chi-square test of independence. Then, we can calculate the test statistic based on the sum of all squared deviations relative to the expected value. The hypotheses can be formulated as

H_0: The distribution is the same for all of the variables

H_0: The distribution is different for the variables

The test statistic can be calculated as

$$Q = \sum_{j=1}^{C} \frac{(\text{observed} - \text{expected})^2}{\text{expected}}$$

Table 10.3 Adults and children buying products A and B

	Product A	*Product B*	
Adults	196	154	350
Children	72	78	150
	268	232	500

Table 10.4 Expected values assuming children and adults have the same distribution

	Product A	*Product B*	
Adults	$350 \cdot 0.536 = 187.6$	$350 \cdot 0.464 = 162.4$	350
Children	$150 \cdot 0.536 = 80.4$	$150 \cdot 0.464 = 69.6$	150
	268 (53.6%)	232 (46.4%)	500

Table 10.5 Squared deviations relative to expected value

	Product A	*Product B*
Adults	$\frac{(196-187.6)^2}{187.6} = 0.376$	$\frac{(154-162.4)^2}{162.4} = 0.434$
Children	$\frac{(72-80.4)^2}{80.4} = 0.878$	$\frac{(78-69.6)^2}{69.6} = 1.014$

where j is the cell number and C is the total number of cells in the table (which in our example is $C = 4$). We can also calculate a table for $\frac{(\text{observed}-\text{expected})^2}{\text{expected})}$ to make it more organized (Table 10.5):

Test statistic is calculated as:

$$Q = \sum_{j=1}^{C} \frac{(\text{observed} - \text{expected})^2}{\text{expected}}$$
$$= 0.376 + 0.434 + 0.878 + 1.014 = 2.702$$

The critical value can be found in the chi-square distribution table with (columns − 1) × (rows − 1) degrees of freedom. In our example, we have columns = 2 and rows = 2 in our table, which results in (columns − 1) × (rows − 1) = (2 − 1) × (2 − 1) = 1 × 1 = 1 degree of freedom. The critical value for a 5% significance level is:

$$cv = \chi^2_{\alpha}(df) = \chi^2_{0.05}(1) = 3.841 \tag{10.3}$$

Hence, we keep the null hypothesis that children and adults have the same distribution.

10.2.1 *Chi-square Goodness of Fit Test*

We may also be interested in testing only one variable. We can test whether a single row or column in a table is different from an expected distribution. This is known as a **chi-square goodness of fit test**, and may be relevant if we are concerned about a specific variable.[1] We then have the hypotheses

H_0: The data is distributed as expected

H_0: The data is not distributed as expected

In the example above, we may want to test if children's behavior is different from the total population. We then have the same test statistic as above

$$Q = \sum_{j=1}^{C} \frac{(\text{observed} - \text{expected})^2}{\text{expected}}$$

where now $C = 2$ since we have only two cells in the row for children. The critical value can be found in the chi-square distribution table with (columns−1) degrees of freedom.

If we want to test if children's preferences for products A and B are different from those of the population in total, we have

$$Q = \sum_{j=1}^{C} \frac{(observed - expected)^2}{expected}$$
$$= \frac{(72 - 80.4)^2}{80.4} + \frac{(78 - 69.6)^2}{69.6} = 0.878 + 1.014 = 1.892$$

The critical value at a 5% significance level is

$$cv = \chi^2_{\alpha}(df) = \chi^2_{0.05}(1) = 3.841 \tag{10.4}$$

where df = columns-1 $= 2 - 1 = 1$. We thus have the same critical value as in the chi-square test above. This is often not the case since we often have more columns than only two. Still, the test statistic is lower than the critical value, so we keep the null hypothesis and conclude that we have not found evidence which allows us to say that children's preferences are different from the population in general.

10.2.2 Gretl and R

In Gretl, you can find the critical value from the chi-squared distribution by going to `Tools > Statistical tables` and selecting the `chi-square` tab. Here, you can enter the degrees of freedom, `df`, and the right-tail probability.

In R, you can find the critical value using the formula `qchisq(alpha, df, lower.tail=FALSE)`, as shown in R code 3.3.

[1] The test when we used the entire table above is often called a chi-square test of independence.

10.2.3 Example with Data

Figure 10.2 shows the dataset in Gretl where we have 500 observations of the dummy variables `child` and `A`.

Manual Calculation

From the given data, we can construct a 2x2 table. But first, we need to perform some additional calculations.

From the data, we have the following:

Variable	Mean	Median	S.D.	Min	Max
child	0.300	0.000	0.459	0.000	1.00
A	0.464	0.000	0.499	0.000	1.00

This means that 30% are children (150 children), and 46.4% (232) of the individuals have purchased item A. Additionally, we also know that 70% are adults (350 adults), and the remaining $500 - 232 = 268$ individuals, or 53.6%, have purchased item B.

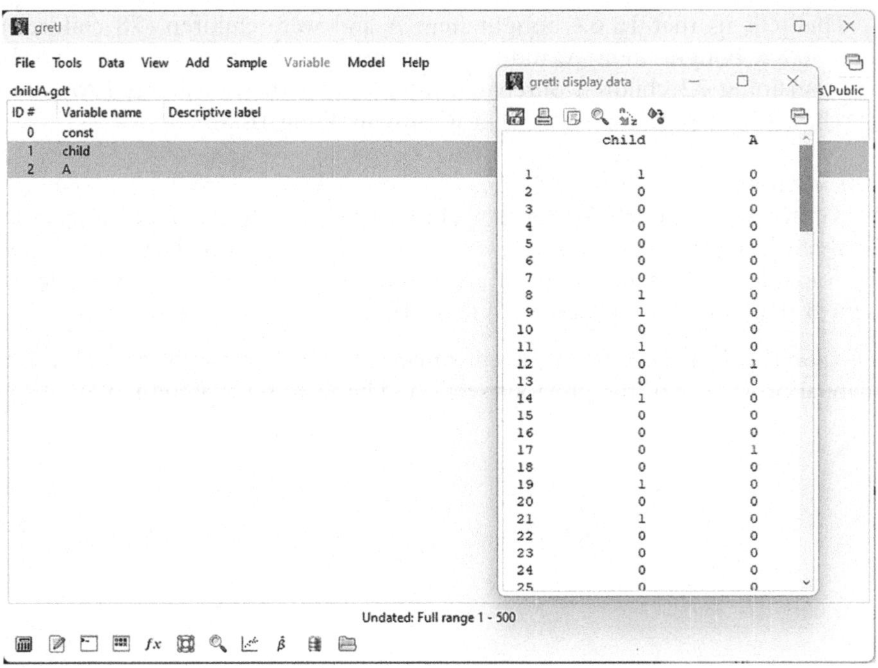

Fig. 10.2 Data for chi-squared test in Gretl

To determine how many children have purchased item A, we can calculate the variable $childA_i = child_i \cdot A_i$ by going to `Add > Define new variable` and entering `childA = child*A`. Furthermore, we can create a dummy variable for adults by going to `Add > Define new variable` and entering `adult = abs(child-1)` (child−1 will yield the value −1 for adults and 0 for children, and taking the absolute value `abs` will convert −1 to 1). Additionally, we can create a dummy variable for adults who have purchased item A by going to `Add > Define new variable` and entering `adultA = adult*A`. This results in the following descriptive statistics:

Summary Statistics, using the observations 1–500

Variable	Mean	Median	S.D.	Min	Max
child	0.300	0.000	0.459	0.000	1.00
A	0.464	0.000	0.499	0.000	1.00
childA	0.156	0.000	0.363	0.000	1.00
adult	0.700	1.00	0.459	0.000	1.00
adultA	0.308	0.000	0.462	0.000	1.00

This tells us that 15.6% bought item A and were children (78 children), and 30.8% bought item A and were adults (154 adults). In other words, the remaining 72 children purchased item B, and the remaining 196 adults purchased item B, in line with what we saw in Table 10.3.

Test in Gretl

In Gretl, we can easily perform a chi-squared test to examine differences between groups by selecting `View > Cross tabulation`. Then, we choose the variables `child` and `A` (or, if applicable, select `adult` and a variable for item B that you may have created) (Fig. 10.3).

Gretl will then perform the chi-squared test automatically, similar to our manual procedure in the previous section. The results are presented as:

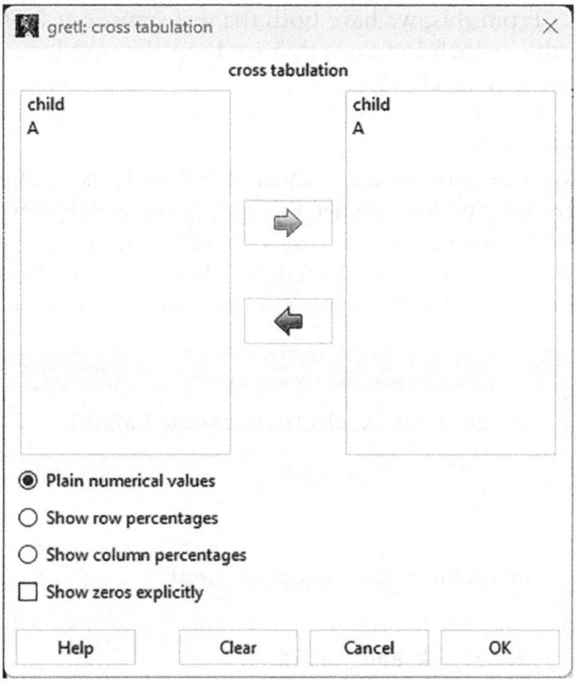

Fig. 10.3 Chi-squared test in Gretl

```
Cross-tabulation of child (rows) against A (columns)

          [   0][   1]  TOT.

[   0]     196   154   350
[   1]      72    78   150

TOTAL      268   232   500

Pearson chi-square test =
     2.70201 (1 df. p-value = 0.100222)


Fisher's Exact Test:
Left:   P-value = 0.959137
Right:  P-value = 0.0611415
2-Tail: P-value = 0.117373
```

From this, we have both the test statistic at 2.702 and the corresponding p-value, which indicates that we keep the null hypothesis for a significance level lower than 10.02%.

Test in R
We can conduct a chi-squared test in R using the `chisq.test()` function. In the function, we need to specify the variables we want to include in the test. Additionally, the function's default setting involves correcting the test results with a continuity correction. Therefore, we also need to include the option `corr = FALSE`. In our example, we use the following code:

R code 10.2: Chi squared test

```
chisq.test(A,child,correct=FALSE)
```

Dette gir

```
Pearson's Chi-squared test


data:  A and child
X-squared = 2.702, df = 1, p-value = 0.1002
```

From this, we see that we get the same test statistic as when we calculated it manually.

10.2.4 Chi-square Test or Regression Analysis

By estimating a logistic regression model (see Sect. 7.6.3) with `A` as the dependent variable and `child` as the explanatory variable, we obtain

$$Pr(A = 1|child) = F(\beta_0 + \beta_1 child)$$

Here, F represents the logistic cumulative distribution function.

Model 1: Logit, using observations 1–500

Dependent variable: A

Standard errors based on Hessian

	Coefficient	Std. Error	z Slope*
const	−0.241162	0.107683	−2.240
child	0.321205	0.195717	1.640800000

Mean dependent var	0.464000	S.D. dependent var	0.499202
McFadden R^2	0.003907	Adjusted R^2	−0.001885
Log-likelihood	−343.9275	Akaike criterion	691.8550
Schwarz criterion	700.2842	Hannan–Quinn	695.1626

*Evaluated at the mean

Number of cases "correctly predicted" = 274 (54.8 percent)
Likelihood ratio test: $\chi^2(1) = 2.698$ [0.1005]

We observe that this leads to a Likelihood Ratio Test using the chi-squared distribution with one degree of freedom. The test statistic and p-value here are the same as those obtained in the chi-squared test, with minor adjustments due to rounding. Conducting such a test within a logistic model is equivalent to performing a multiple hypothesis test on the dependent variables to assess whether any of the independent variables has a significant effect on the dependent variable. In this case, with the dependent variable being product A and the sole independent variable being the dummy variable for child, this test evaluates whether being a child or an adult influences the purchase of the product.

$$H_0\colon \beta_1 = 0$$
$$H_A\colon \beta_1 \neq 0$$

It is worth noting that we retain the null hypothesis at a 5% significance level.

Chi-squared Test or Logistic Model

Conducting a multiple hypothesis test in a logistic model where the null hypothesis is that none of the explanatory variables have a significant effect on the dependent variable is equivalent to conducting a chi-squared test.

Table 10.6 Wages and Country of Residence

Wage	*City*	*France*	*Switzerland*	*Germany*
320	301	1	0	0
290	5001	0	0	1
170	301	1	0	0
480	5001	0	0	1
510	4601	0	1	0
410	5001	0	0	1
650	4601	0	1	0
370	301	1	0	0

10.3 Analysis of Variance

If we wish to compare more than two groups (as we did with a t-test in this chapter), we can employ analysis of variance. For instance, we can group the eight individuals used as an example based on the country they live in, if we are interested in wage differences between countries. As illustrated in Table 10.6, three individuals lived in France, two in Switzerland, and three in Germany.

We can group these individuals based on the city and wage as shown in Table 10.7:

We can observe wage differences among the groups. For example, the mean wages in the three countries are as follows:

$$\bar{Y}_1 = \frac{\sum_{i=1}^{n} Y_{1,i}}{n_1} = \frac{320 + 170 + 370}{3} = 286.67$$
$$\bar{Y}_2 = \frac{\sum_{i=1}^{n} Y_{2,i}}{n_2} = \frac{510 + 650}{2} = 580.00$$
$$\bar{Y}_3 = \frac{\sum_{i=1}^{n} Y_{3,i}}{n_3} = \frac{290 + 480 + 410}{3} = 393.33$$

Here, subscript 1 refers to France, 2 to Switzerland, and 3 to Germany. We observe that the mean wage is highest in Switzerland and lowest in France. However, there is also considerable variation within each country. This variation introduces uncertainty when using the sample means for each country as

Table 10.7 Wages and City of Residence

France	*Switzerland*	*Germany*
320	510	290
170	650	480
370		410

estimates of the population means for the countries. To test if the wage difference between the countries is significant, we can use **analysis of variance**. In this analysis, we account for the variation within each group (country) and compare the mean for each group (country) to the overall mean value. Although we observe differences in mean wages between the countries, we must consider the variation as well. Therefore, we need to compare the variation between groups (how the mean for each country compares to the overall mean value) with the variation within each group. This can be set up as a hypothesis test.

$$H_0\colon \mu_1 = \mu_2 = \cdots = \mu_j$$
$$H_A\colon \text{At least two population means are different}$$

For j different groups. In our example, we use:

$$H_0\colon \mu_1 = \mu_2 = \mu_3$$
$$H_A\colon \text{At least two population means are different}$$

This hypothesis test can be conducted using an F-test, as explained in Sect. 6.4. The test statistic is given by:

$$F = \frac{S^2_{\text{between}}}{S^2_{\text{within}}}$$

where S^2_{between} is the variance between groups, and S^2_{within} is the variance within groups. These can be calculated as follows:

$$S^2_{\text{between}} = \sum_{j=1}^{g} n_j(\bar{Y}_j - \bar{Y})^2/(g-1)$$

$$S^2_{\text{within}} = \sum_{i=1}^{n} (Y_i - \bar{Y}_j)^2/(n-g)$$

where j is the group number and g is the number of groups. In our example, we obtain:

$$\begin{aligned} S^2_{\text{between}} &= \sum_{j=1}^{g} n_j(\bar{Y}_j - \bar{Y})/(g-1) \\ &= [3 \cdot (286.67 - 400)^2 + 2 \cdot (580.00 - 400)^2 + 3 \cdot (393.33 - 400)^2]/2 \\ &= 103464.53/2 = 51733.27 \end{aligned}$$

and

$$S^2_{\text{within}} = \left[\sum_{i=1}^{n} (Y_i - \bar{Y}_j)^2\right]/(n-g)$$

$$
\begin{aligned}
&= [(320 - 286.67)^2 + (170 - 286.67)^2 + (370 - 286.67)^2 \\
&\quad + (510 - 580.00)^2 + (650 - 580.00)^2 \\
&\quad + (290 - 393.33)^2 + (480 - 393.33)^2 + (410 - 393.33)^2]/(8 - 3) \\
&= 49933.33/5 = 9986.67
\end{aligned}
$$

The test statistic is therefore

$$F = \frac{S^2_{\text{between}}}{S^2_{\text{within}}} = \frac{51733.27}{9986.67} = 5.18$$

We can compare this with the critical value using $g - 1$ degrees of freedom in the numerator and $n - g$ degrees of freedom in the denominator. In this case, there are $g - 1 = 3 - 1 = 2$ degrees of freedom in the numerator and $n - g = 8 - 3 = 5$ degrees of freedom in the denominator. This results in a critical value at a 5% significance level of

$$cv = F_\alpha(g - 1, n - g) = F_{0.05}(2, 5) = 5.786$$

Since the test statistic is less than the critical value, we choose to retain the null hypothesis that the group means are equal. Therefore, the mean wages are not significantly different in the three cities.

10.3.1 Example in Gretl

In order to perform analysis of variance in Gretl, we can go to `Model > Other linear models > ANOVA`. We can then choose `wage` as `Response variable` (dependent variable) and `country` as `Treatment variable`.

We then get

```
Analysis of Variance, response = wage, treatment = country:

             Sum of squares  df     Mean square


Treatment          103467   2         51733.3
Residual          49933.3   5         9986.67
Total              153400   7         21914.3

F(2, 5) = 51733.3 / 9986.67 = 5.18024 [p-value 0.0605]

Level          n        mean      std. dev

301            3     286.667       104.08
4601           2         580       98.995
5001           3     393.333       96.090

Grand mean = 400
```

We find the test statistic to be 5.18 and the p-value to be 0.0605. This shows that at a 5% significance level, we keep the null hypothesis that the population means are equal.

10.3.2 Example in R

In R, we can use the function `aov(depvar ~ explvar)`, where `depvar` is the dependent variable and `explvar` is the explanatory variable, in order to do analysis of variance, and then we can look at the summary of this. The explanatory variable needs to be a factor variable in R. If it is not, we can add `factor()` for this variable.

In our example, we thus use

R code 10.3: Analysis of variance

```
anova <- aov(wage ~ factor(country))
summary(anova)
```

We then get

```
                Df Sum Sq Mean Sq F value Pr(>F)
factor(country)  2 103467   51733    5.18 0.0605 .
Residuals        5  49933    9987
---
Signif. codes:
    0 '***' 0.001 '**' 0.01 '*' 0.05 '.' 0.1 ' ' 1
```

This gives the test statistic 5.18 and the p-value 0.0605, in line with what we calculated manually.

10.3.3 Analysis of Variance or Using Dummy Variables

Analysis of variance is carried out through a hypothesis test as an F-test where we test if population means are equal. This can also be done in the regression model if we have dummy variables for fixed effects. Fixed effects will also measure differences in means, so testing whether these dummy variables have a significant effect on the overall mean will be equivalent to the analysis of variance here. The test statistic of 5.18 is in line with the test statistic we got from the multiple hypothesis test after estimating (7.1).

Analysis of Variance or Use of Dummy Variables

The analysis of variance as presented here can also be tested by a multiple hypothesis test on dummy variables as explanatory variables in a regression model.

Bibliography

Akaike, H. (1974). A new look at the statistical model identification. *IEEE Transactions on Automatic Control, 19*(6), 716–723.

Alexander, N. (2015). What's more general than a whole population? *Emerging Themes in Epidemiology, 12*, 1–5.

Bårdsen, G., & Lütkepohl, H. (2011). Forecasting levels of log variables in vector autoregressions. *International Journal of Forecasting, 27*(4), 1108–1115.

Breusch, T. S., & Pagan, A. R. (1979). A simple test for heteroscedasticity and random coefficient variation. *Econometrica: Journal of the Econometric Society, 47*(5), 1287–1294.

Card, D. (1995). Using geographic variation in college proximity to estimate the return to schooling. In L. N. Christophides, E. K. Grant, & R. Swidinsky (Eds.), *Aspects of labour market behavior: Essays in honor of John Vanderkamp* (pp. 201–222). Toronto: University of Toronto Press.

Engle, R. F., & Granger, C. W. (1987). Co-integration and error correction: Representation, estimation, and testing. *Econometrica: Journal of the Econometric Society, 55*(2), 251-276.

Greene, W. H. (2003). *Econometric analysis*. Pearson Education.

Hannan, E. J., & Quinn, B. G. (1979). The determination of the order of an autoregression. *Journal of the Royal Statistical Society: Series B (Methodological), 41*(2), 190–195.

Johnson, S. R. (2008). The trouble with QSAR (or how I learned to stop worrying and embrace fallacy). *Journal of Chemical Information and Modeling, 48*(1), 25–26.

King, G., & Roberts, M. E. (2015). How robust standard errors expose methodological problems they do not fix, and what to do about it. *Political Analysis, 23*(2), 159–179.

Koenker, R. (1981). A note on studentizing a test for heteroscedasticity. *Journal of Econometrics, 17*(1), 107–112.

Koop, G. (2013). *Analysis of economic data*. Wiley.

Schwarz, G. (1978). Estimating the dimension of a model. *Annals of Statistics, 6*(2), 461–464.

Vannette, D. L., & Krosnick, J. A. (Eds.). (2018). *The Palgrave handbook of survey research*. Springer International.

B. Karlsen Kivedal, *Applied Statistics and Econometrics*,
https://doi.org/10.1007/978-3-031-53142-2

White, H. (1980). A heteroskedasticity-consistent covariance matrix estimator and a direct test for heteroskedasticity. *Econometrica: Journal of the Econometric Society, 48*(4), 817–838.

Wooldridge, J. (2019). *Introductory econometrics: A modern approach*. Cengage Learning.

Index

A
Adjusted R-squared, 75
All else being equal, 98
ARIMA model, 198
Augmented Dickey-Fuller test, 200
Autocorrelation, 193
Average, 24

B
Best linear unbiased estimators (BLUE), 85
Breusch-Godfrey test, 214
Breusch-Pagan test, 110

C
Causal effect, 116
Ceteris paribus, 97
Chow test, 217
Cointegrated, 220
Cointegration, 220
Confidence interval, 27
Confidence interval of a coefficient, 78
Correlation, 29
Covariance, 28
Critical value, 41

D
Degrees of freedom, 25
Derivative, 61, 170
Descriptive statistics, 19
Deterministic trend, 219
Deviation, 62
Dickey-Fuller test, 199
Drift, 205
Dummy variables, 125
Durbin-Wu-Hausman test, 116

E
Endogeneity, 115
Engle-Granger test, 220
Error correction model, 220
Error term, 62, 71
Expectations operator, 23

F
Fixed effect, 131
Forecast error, 209

G
Gdt file, 7

H
Hausman test, 116
Heterogeneity, 85
Heteroskedasticity, 84, 110
Homoskedasticity, 84

I
Information criteria, 197

B. Karlsen Kivedal, *Applied Statistics and Econometrics*,
https://doi.org/10.1007/978-3-031-53142-2

Instrumental variable, 115
Interaction effects, 134
Intersection point, 60
Interval estimate, 27
Isolated effect, 97

L
Law of large numbers, 34
Lin-log model, 161
Logarithm, 161
Logit model, 151
Log-lin model, 164
Log-log model, 165

M
Maximum likelihood, 199
Mean, 24
Median, 24
Mode, 24
Model, 62
Multicollinearity, perfect, 126
Multiple hypothesis testing, 106

N
Non-linear effect, 22
Non-linear model, 158
Normally distributed error term, 84

O
Observation range, 65
Omitted variable bias, 122
Ordinary least squares method (OLS), 59
Overfitted, 114

P
Parameter stability, 216
Partial derivative, 97
Partial effect, 23
Perfect multicollinearity, 126
Polynomial function, 166
Posit, 12
Prediction error, 70
Probaility distributions, 31
Proxy variable, 115
P-value, 56

R
Regression line, 59
Residual, 62, 71
Robust standard errors, 113
R-squared, 74

S
Sample, 19
Script, 14
Secondary data source, 20
Semi-logarithmic model, 161, 164
Shift dummy, 217
Significance level, 41
Slope coefficient, 21, 60
Spurious relationship, 186
Standard error, 77
Standard normal distribution, 32
Stationary, 193
Statistical inference, 19
Stochastic trend, 220

T
T-distribution, 34
Time series data, 180
Trend, 185
Trend-stationary, 200
T-test, paired, 53
Two-stage least squares, 115

U
Unbiased, 84
Unbiased estimates, 121
Uniform probability distribution, 33

W
White test, 110

The manufacturer's authorised representative in the EU is Springer Nature Customer Service Centre GmbH, Europaplatz 3, 69115 Heidelberg, Germany. If you have any concerns regarding our products, please contact ProductSafety@springernature.com

Printed and bound by CPI Group (UK) Ltd, Croydon, CR0 4YY
28/09/2026
02224087-0003